BIOCOMPUTING

Biocomputing

Volume 1

Biocomputing

Edited by

Panos M. Pardalos
University of Florida,
Gainesville, Florida, U.S.A.

and

Jose Principe
University of Florida,
Gainesville, Florida, U.S.A.

KLUWER ACADEMIC PUBLISHERS
DORDRECHT / BOSTON / LONDON

A C.I.P. Catalogue record for this book is available from the Library of Congress.

ISBN 1-4020-0641-1

Published by Kluwer Academic Publishers,
P.O. Box 17, 3300 AA Dordrecht, The Netherlands.

Sold and distributed in North, Central and South America
by Kluwer Academic Publishers,
101 Philip Drive, Norwell, MA 02061, U.S.A.

In all other countries, sold and distributed
by Kluwer Academic Publishers,
P.O. Box 322, 3300 AH Dordrecht, The Netherlands.

Printed on acid-free paper

Printed in the Netherlands.

Contents

Preface

In the quest to understand and model the healthy or sick human body, researchers and medical doctors are utilizing more and more quantitative tools and techniques. This trend is pushing the envelope of a new field we call Biomedical Computing, as an exciting frontier among signal processing, pattern recognition, optimization, nonlinear dynamics, computer science and biology, chemistry and medicine.

A conference on *Biocomputing* was held during February 25–27, 2001 at the University of Florida. The conference was sponsored by the Center for Applied Optimization, the Computational Neuroengineering Center, the Biomedical Engineering Program (through a Whitaker Foundation grant), the Brain Institute, the School of Engineering, and the University of Florida Research & Graduate Programs. The conference provided a forum for researchers to discuss and present new directions in Biocomputing. The well-attended three days event was highlighted by the presence of top researchers in the field who presented their work in Biocomputing. This volume contains a selective collection of refereed papers based on talks presented at this conference. You will find seminal contributions in genomics, global optimization, computational neuroscience, FMRI, brain dynamics, epileptic seizure prediction and cancer diagnostics.

We would like to take the opportunity to thank the sponsors, the authors of the papers, the anonymous referees, and Kluwer Academic Publishers for making the conference successful and the publication of this volume possible.

Panos M. Pardalos and Jose C. Principe
University of Florida
January 2002

Chapter 1

MAKING SENSE OF BRAIN WAVES: THE MOST BAFFLING FRONTIER IN NEUROSCIENCE

Walter J. Freeman

Department of Molecular & Cell Biology,
University of California at Berkeley
wfreeman@socrates.berkeley.edu

Abstract   Brains are characterized by every property that engineers and computer scientists detest and avoid. They are chaotic, unstable, nonlinear, nonstationary, non-Gaussian, asynchronous, noisy, and unpredictable in fine grain, yet undeniably they are among the most successful devices that a billion years of evolution has produced. No one can justifiably claim that he or she has modeled brains, but they are a flowing spring of new concepts, and they provide a gold standard of what we can aspire to accomplish in developing more intelligent machines. The most fertile source of ideas with which to challenge and break the restrictions that characterize modern engineering practice is the electroencephalogram (EEG). It was the action potential of single neurons that provided the foundation of neurobiology for the 20th century, and in its time it supported the development of digital computers, neural networks, and computational neuroscience. Now in the 21st century, the EEG will lead us in a remarkably different direction of growth for the computing industry, which will be dominated by highly parallel, hierarchically organized, distributed analog machines. These devices now exist in prototype form. They feed on noise in support of chaotic attractor landscapes, which are shaped by reinforcement learning through self-governed experience, not training by 'teachers', and they may solve many of the problems of interfacing between finite state automata and the infinite complexity of the real world.

1.   Introduction

Electroencephalographic (EEG) potentials are recorded from the scalp as aperiodic fluctuations in the microvolt range in humans and animals. They are also found within brains where they are often referred to as 'local field potentials', but they are manifestations of the same dynamics of populations of neurons. This report summarizes how they arise, why they oscillate, and what

1

P.M. Pardalos and J. Principe (eds.), Biocomputing, 1-23.
© 2002 *Kluwer Academic Publishers. Printed in the Netherlands.*

they can tell us about brain function that will be useful for engineers to construct devices that can perform some of the functions that brains do very well, such as pattern recognition and classification, in contrast to existing machines.

Neurons have two kinds of filaments by which they interconnect and interact. The transmitting filament, the axon, generates propagating pulses called action potentials or 'units', which serve to communicate information over short and long distances without attenuation. The receiving filament, the dendrite, converts pulses to graded waves of current that can be summed linearly both in time and space. The resultant sum is re-converted to pulses at rates proportional to dendritic current density in a biological form of pulse frequency modulation. Dendrites consume 95 % of the energy that brains use for information processing, axons only 5 %, so they are the principal determinants of the patterns that are observed in brain imaging with fMRI, PET and SPECT. They are also the principal source of electric currents passing across the extraneuronal resistance of brain tissue (300 ohm.cm/cm2), creating the EEG. Most of what we know about sensory and motor information processing has come from studies of action potentials. The bridge that we need to understand the relations between 'unit activity' and brain images is provided by the EEG.

2.   Relations between EEG and 'units'

Most EEG waves are generated by the cerebral cortex of the forebrain, for two reasons. First, the neurons are exceedingly numerous, and they are organized in layers with their dendrites in the main oriented parallel to each other (Figure 1.1), so their currents are also aligned in summation. Second, they interact synaptically in sparse but high density feedback. The reciprocal synaptic connections are by both positive feedback (mutually excitatory) leading to large areas of spatially coherent activity, and negative feedback (between excitatory and inhibitory populations) leading to oscillations in the gamma range (20-80 Hz) that carry information content in wave packets [12] Multiple types of feedback (including positive inhibitory feedback leading to spatial contrast enhancement) support lower frequency oscillations in alpha and theta ranges (3-12 Hz) at which wave packets are gated.

To understand how the information that is received at the dendrites is converged to the axon we need to understand the way the dendrites operate. The dendrites typically generate electric currents which are initiated at synapses packed over the entire tree (Figure 1.2).

Each synapse acts like a small battery with a high internal impedance, so that its current strength does not vary with external load. By Kirchoff's law, current always flows in a closed loop, so that current must flow in one direction across the membrane at the synapse and in the opposite direction across the membrane at other sites with impedance match to the crossing. The preferred

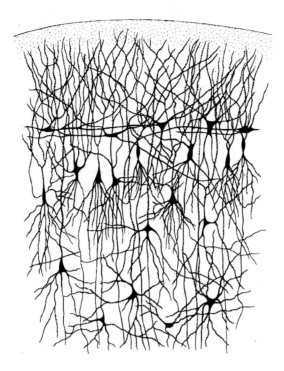

Figure 1.1. The filamentous structure of cortical neurons and their palisades in layers is shown by silver impregnation (the Golgi technique), selecting only 1 % of the neurons in the field

current path is along the dendritic shaft toward the cell body. The loop currents from all synapses superimpose and are converge to the origin of all the branches at the cell body and pass across the membrane of the initial segment of the axon where it originates from the cell body. Functionally, the initial segment of the axon is also known as the trigger zone.

Among the three kinds of chemical synapse, an excitatory synapse causes current to flow inwardly at the synapse, along the dendritic cable away from the synapse, outwardly across the membrane, and back to the synapse in the space outside the membrane (Figure 1.2). An inhibitory synapse causes a current loop with flow outwardly at the synapse and inwardly everywhere else. Each current causes a voltage drop across a high resistance in the trigger zone membrane. The inhibitory sum is subtracted from the excitatory sum of voltage differences.

The inflow of current at one end of the neuron and the outflow at the other end create a source-sink pair, which gives a dipole field of potential. The amplitude falls roughly with the square of distance from the center, which explains how the EEG can be observed at the surface of the cortex and scalp, though at one

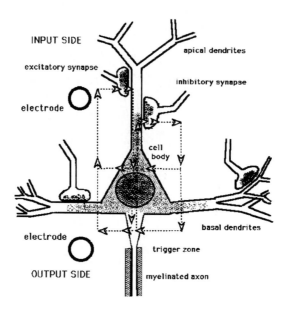

Figure 1.2. Dendrites receive action potentials at synapses and convert the impulses to loop currents. The sum of current determines the rate of unit firing. The same current contributes to the EEG recorded with transcortical electrodes, but it sums with the current from all neurons in the neighborhood, manifesting a local mean field.

thousandth the amplitude of transmembrane potentials, owing to the low specific resistance of extraneuronal fluid compared to that of the neural membrane.

More generally, because the spread and summation of dendritic current conform to the principle of superposition, they are described by a 2nd order linear ordinary differential equation, which is evaluated by fitting the sum of two exponentials to the impulse response.

$$d^2v/dt^2 + (a+b)dv/dt + abv = k_{ij}G(p) \qquad (1)$$

The minimal input to an excitatory synapse is an action potential, which is an impulse lasting about one millisecond. With a single impulse at the synapse the dendritic current rises at a rate of $1/b = 1$ to 3 milliseconds and decays at a rate of $1/a = 5$ to 10 milliseconds. At the cell body and trigger zone this brief electrical event is observed as an excitatory postsynaptic potential (EPSP). The impulse response of an inhibitory synapse has the similar form and decay time but opposite polarity and is called an inhibitory postsynaptic potential (IPSP).

These two types of linear response are modeled in hardware with operational amplifiers and (for inhibitory synapses) inverters, and the combinations of multiple synapses are modeled with summing junctions. Summing at trigger zones takes place after synaptic amplification and sign reversal.

The third type of synapse, which is called modulatory, does not induce a loop current. Instead, it changes the strength of action of adjacent excitatory and inhibitory synapses. Neuromodulators are also released to diffuse through populations and enable alterations of their chemical states. Such multiplicative actions are modeled with variable resistors or a variable gain amplifier in circuits using time multiplexing to solve the connectivity problem (Figure 1.3).

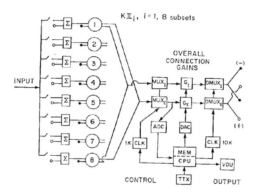

Figure 1.3.   Time multiplexing is feasible because the bandwidth of EEG is much lower than that of 'units'. This reduces connectivity to 2N instead of N**2 for N nodes. Gain values are indexed to connection pairs under digital control. Analog amplitudes simulate pulse density coding.

Neuromodulation is used to normalize neural activity levels across distributed input, to perform logarithmic conversion and range compression on input, to change overall neuronal gain in respect to behavioral awakening and arousal, and to enable selective changes in synaptic strength during learning, either an increase in sensitivity during association or a decrease during habituation.

The architecture of dendrites reflects a key role that the dendrites perform, which is to convert incoming pulses to currents and sum them at the initial segment. Spatial integration at the cell body is important, simply because that is where all the axonal and dendritic branches originate. The integration of the dendritic currents is done there also across time. When stimuli are frequent enough, new current superimposes on old fading current to give temporal summation at the initial segment. The dendritic branches are poor electrical conductors with high internal resistance and high leakage across their surfaces. Effective dendritic currents can seldom be transmitted passively over distances longer than a few hundred microns because of the attenuation with distance. The critical function of the axon is to transmit the time-varying amplitude of current summed by the dendrites to distant targets without attenuation.

The neuronal stratagem is to convert the sum of currents into a pulse train, in which pulse frequency is proportional to wave amplitude. The pulse train is converted back to a wave function that is smoothed by the target dendrites. This is why the axonal "signal" is "analog" and not "digital". The energy for transmission along the axon is provided locally by the axon membrane as the pulse propagates, but the release of the energy takes time It is not done with the speed of an electronic conductor. The price for axonal output that is free of attenuation and can be carried over large distances is threefold: nonlinearity; a delay in pulse transmission; and discretization in time. Thus dendrites which are short do not use pulses, and axons which are long do. Dendrites generate graded currents that are superimposable and distributed in time and space, whereas the pulses of axons are not superimposable. They are point processes in time and space.

In a neural population, activity densities are defined over spatial distributions of neurons in the two surface dimensions of the cortex orthogonal to the orientation of the dendrites. The transformation at trigger zones of dendritic wave density to axonal pulse density (Figure 1.4) has bilateral saturation. The function for single neurons is linear and time-varying; for populations it is nonlinear and static. Only populations of neurons have the smooth static "sigmoid" relationship between the two state variables. The reason that the curve is static stems from the fact that individual neurons in the population fire unpredictably, as shown by their Poisson interval distributions. They also fire in an uncorrelated manner with respect to each other. Hence the refractory period and adaptation of the single neuron do not appear directly as time variance in the ensemble average of activity of populations but set the upper limit on normalized pulse density, $Q_m = (P - P_o)/P_{o'}$.

Experimental demonstration of the sigmoid curve (Freeman 2000) is by calculating the pulse probability conditional on EEG amplitude in the same population. Cortical populations display activity when without stimulation that is seen as "spontaneous" or "background" activity at rest. The pulse activity of any one neuron averaged over a long time period is assumed by the ergodic hypothesis to conform to the activity of its population over a brief time period. Its mean rate serves to represent the mean pulse density, P_o (Figure 1.4). In an excited population the pulse density increases up to a limit P_m, which is determined largely by potassium conductance, G_K, at trigger zones. In an inhibited population (hyperpolarization) the pulse density is bounded at zero. In awake subjects the ratio $Q_m = P_m/P_o$ ranges from 5:1 to 12:1. In resting and sleeping states the ratio drops to 2:1, and under deep anesthesia it goes to zero, giving the open loop state for the impulse response.

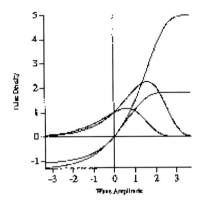

Figure 1.4. The asymmetric sigmoid curve is shown for two values of the asymptotic maximum, Q_m, the lower at 1.8 for rest, the upper at 5.0 for motivation and arousal as shown by the bursts in Figure 1.3. The equations are:

$$p = p_o(1 + \{1 - \exp[-(e^v - 1)/Q_m]\}), v > -u_o \tag{2}$$

$$dp/dv = u_o \exp[v - (e^v - 1)/Q_m] \tag{3}$$

Another characteristic of the trigger zones is that P_o, Q_m and the slope dp/dv of the sigmoid curve (Figure 1.4) increase with behavioral arousal of a subject. This state dependence is expressed by a single variable Q_m in normalized coordinates, which probably reflects the operation on the cortex of a single neuromodulator chemical that is responsible for regulating cortical excitability during arousal in various behavioral states. This property is part of a more general pattern, in which each area of cortex receives modulatory input from several other parts of the brain. The modulation does not provide information-specific input, but it changes and adjusts the cortical state in such ways as "turn on", "turn off", "attend", "learn", "habituate", and so on by simultaneously operating on populations of cortical neurons. Neuromodulators may best be simulated in VLSI embodiments by sending control signals through the power lines of diodes and amplifiers.

The most important dynamic aspect of the biological sigmoid curve is the fact that it is asymmetric. Its maximal slope dp/dv is displaced to the excitatory side. This property reflects the fact that in a population most of the neurons most of the time operate near equilibrium and just below their thresholds for firing, where voltage-dependent G_{Na} increases exponentially with depolarization. As a result of this regenerative feedback the firing probability increases exponentially with depolarization, and this is reflected in the exponential increase of the concave-upward part of the sigmoid curve. The slope dp/dv is a main determinant of the forward gain of a population Therefore, the gain is both state-dependent (on the degree of arousal) and input-dependent, giving a strong increase in gain on sensory excitation of the population. These gain dependencies are crucial for state transitions of cortex during behavioral information processing. In particular, the nonlinear gain renders local areas of cortex unstable in respect to input, and the instability is enhanced with increasing arousal as in fear and anger.

3.   Three levels of hierarchical coding

There are three main levels of neural function in the pattern recognition performed by sensory systems, which are to be modeled with appropriately tailored state variables and operations. Microscopic activity is seen in the fraction of the variance of single neuron pulse trains (>99.9 %) that can be correlated with sensory input in the form of feature detector neurons and motor output in the form of command neurons. Mesoscopic activity [20] is carried in the <0.1 % of the total variance of each neuron that is covariant with other neurons in neuropil that comprise local neighborhoods. Collectively it is observed in dendritic potentials (EEGs). The interaction of multiple sensory cortices in different modalities together with association areas constitutes a macroscopic system [19]. That is the hierarchical level of organization of brain function that is revealed by present-day devices for brain imaging that rely on metabolic activity and blood flow. The interactions by local negative feedback between excitatory and inhibitory neurons at the mesoscopic level support periodic oscillations manifesting limit cycle attractors. Multiple areas of neuropil comprising a sensory system interact by long feedback paths. Due to the facts that locally they have incommensurate characteristic frequencies, and that multiple local areas interact by axonal transmission imposing relatively long delays, together they maintain global chaotic states that are observed in aperiodic oscillations. The basal state without stimulus input is characterized by Gaussian distributions of EEG amplitude and by power spectral densities that conform to $1/f^2$ linear decrease in log power with log frequency (brown noise).

On destabilization by sensory input the multipart system undergoes a state transition to a more narrowly constrained spectral distribution in the gamma range, which results from substantial increase in interactions of cortical neurons with each other, so that they no longer accept internal input. The neurons create a new pattern of activity following each sensory barrage (Figure 1.5). The contents of neural activity in brains are deduced from correlations of the amplitudes and frequencies of the observed activity with the on-going behavior of animals. These correlations reveal a *profound difference between the dynamics of sensation and that of perception.*

Mesoscopic activity in sensory cortices is perceptual, not sensory. EEGs of nonlinear brain dynamics manifest internal self-organizing dynamics of newly

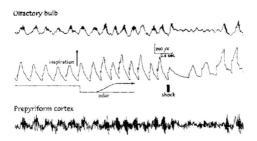

Figure 1.5. An example from the olfactory bulb shows the EEG manifesting the short bursts of activity (top trace) constituting wave packets, the formation of which is by destabilization of the bulb by input, owing to the asymmetric nonlinear gain shown in Figure 4. The middle trace shows the pattern of driving by respiration that is under control by the limbic system. The bottom EEG is from the target of bulbar transmission.

created patterns emerging from cortical background noise destabilized under perturbation by sensory stimuli. Sensory and perceptual contents coexist in cortices. The former are extracted by time ensemble averaging over trials, the latter by spatial ensemble averaging of multichannel simultaneous recordings on single trials [3, 12, 15, 20].

The contents of brain activity patterns are conveyed by wave packets of oscillatory activity recurring at rates in the theta range having the same aperiodic wave form observed throughout the entire population (Figure 1.6, left frame).

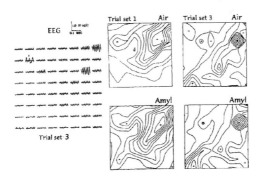

Figure 1.6. The left frame shows the carrier wave form of a single wave packet. The right frames show the changes in AM pattern with classical conditioning. The visual, auditory and somatic cortices have the same form of coding: a chaotic carrier wave spatially amplitude-modulated.

This spatially coherent wave serves as a carrier for content by amplitude modulation in space (right frames). Because the AM patterns are created by interactions of neurons by synapses that have been modified by experience, patterns are not invariant with stimuli but reflect instead that experience. Hence the internal context of cortical responses to stimuli is carried in the spatial domain of cortical dynamics by spatial AM of the aperiodic carrier wave form [28].

During behavioral conditioning of an animal a new AM pattern emerges with each new stimulus the animal learns to discriminate[26], implying that the sensory system maintains an attractor landscape with a basin for each class.

The basin of each attractor is defined by the stimuli in the learning set that is paired with reinforcement. Each basin is accessed by a state transition that is driven by a surge of input from sensory receptors, because the entire landscape is brought into play, then suppressed during exhalation to enable a new sample to be taken (Figure 1.7). The basin allows for many-to-one convergence that takes place when the animal generalizes over samples to identify the class to which a stimulus belongs. An important feature is that the chaos is a mesoscopic property. The microscopic neurons are governed by point attractors, not by chaotic attractors.

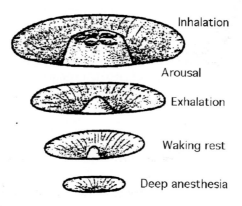

Figure 1.7.   The state space of the olfactory system is schematized as it might appear from a high dimensional space projected into 2-space. The attractor landscape appears with inhalation, and it dissolves with exhalation to allow the next stimulus to be classified.

The state transition is manifested by a conic phase gradient (Figure 1.8) of the aperiodic oscillatory event [21]. The apex of the cone demarcates the site of nucleation for each state transition, the location and sign of which vary randomly. The random variation in sign (extreme lead or lag) shows that the apex cannot represent the location of a pacemaker. The phase gradient shows the group velocity by which it spreads [21, 22, 24].

Phase cones play a major role in determining the size of wave packets in neocortex, which is a continuous sheet of neuropil over each entire cerebral

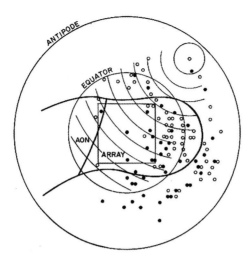

Figure 1.8. Phase distributions were measured with respect to the phase of the spatial ensemble average at the surface of the olfactory bulb and fitted with a cone in spherical coordinates. The sketch is a projection of the outline of the bulb as it would appear on looking through the left bulb onto the array an the lateral surface of the bulb. A representative set of isophase contours is at intervals of 0.25 radians/mm. The locations of the apices of the cones on the surface of the sphere (2.5 mm in radius) are plotted from the center of the array to the antipode. The square outlines the electrode array. The standard error of location of points was twice the radius of the dots.

hemisphere. The delay that is imposed in the state transition from the site of nucleation by the conduction velocities of the axons that run parallel to the surface causes progressive phase difference with distance, eventually to the extent of going out of phase. This offers a solution to the problem of how differing wave forms coexist in the same hemisphere. The phase gradient creates a soft boundary condition that is approximated by the half-power radius, implying that the modal diameter of neocortical wave packets is about a centimeter [23].

These two kinds of coding coexist in brains. The *microscopic* coding is characteristic of single neurons in the brain stem and spinal cord as well as in the peripheral nervous system, but also of cortical neurons. The microscopic pulse trains are described by means of stochastic point processes. The *mesoscopic* coding is found in the cortices as local mean fields of activity that constitute averages over many neurons. This activity has continuous distributions of neural activity, with time and space constants 1 to 2 orders of magnitude larger than those for single neurons. Though it appears random in time, this activity is spatially coherent and highly structured in phase and amplitude [17, 18]. Hence equations for describing neurodynamics have two forms: discrete and stochastic difference equations to model the microscopic input and output channels of cerebral cortex, and continuous integrodifferential equations for intracortical dynamics [12, 15]. While neural networks can be modeled with matrices that represent the dynamics of local integrate-and-fire elements connected globally, whether fully or sparsely, the dynamics of cerebral cortex is modeled with arrays of coupled oscillators in two spatial dimensions, with sparse but global internal connectivity to represent the architecture of the neural populations of cortex.

The distinction between digital and analog embodiments is focused on the difference between representing state variables with numbers versus voltages, not with discrete versus continuous variables. Brain function is neither analog nor digital, as these terms are defined for computer usage. Pulse trains that appear to be digital are in fact analog as a form of pulse frequency modulation. Analog integration is done with continuous variables in time, but usually with discretization by compartments in space and for segmenting for multiplexing (Figure 1.3) to solve the connectivity problem [11]. Sums of dendritic current are locally continuous distributions in time and space for short segments of time, but their spatial patterns are discretized by discontinuities imposed by 1st order state transitions (Figure 1.5) to form wave packets [12, 15, 16]. New brain models will be hybrid, not analog or digital.

Parallel networks of coupled oscillators in software [36] and hardware [11] serve to model the nonlinear dynamics of cortex. Solutions of ordinary differential or difference equations serve to model the chaotic wave forms of normal and abnormal brain activity, including the spatial coherence of broad-spectrum aperiodic activity that is so characteristic of EEG fields of potential over the cerebral cortex. The spatially coherent mesoscopic activity is extracted by the targets of cortical transmission through divergent-convergent axonal tracts that perform spatial integral transformations on cortical outputs. By this operation the cooperative, spatially coherent chaotic activity is extracted as signal, and the input-driven, spatially incoherent activity is attenuated as noise [20].

4.   Simulation of "background" activity

Unlike most neural network models, which remain at rest until given input, the cerebral cortex is ceaselessly active in sleep and wakefulness. The key to understanding the hierarchical organization of brain function lies in explaining the dynamical origin of this sustained endogenous activity. At the microscopic level, the background activity of single neurons in the absence of sensory stimulation can be accounted for by relaxation oscillation, which tends to give periodic or quasiperiodic pulse trains manifesting one or more limit cycle attractors. which is consistent with the performance of the "integrate and fire" models of single neurons.

In contrast, neurons embedded in neuropil typically generate aperiodic pulse trains [1, 2, 12, 24]. Their autocorrelations reveal their refractory periods but seldom clear characteristic frequencies. Their interval histograms conform to the Gamma distribution of order $1/2$ at modest mean rates, tending toward the Poisson distribution (with a dead time) at low rates and the Gaussian distribution at high rates. Crosscorrelations between adjacent neurons are vanishingly small.

The background activity of neuropil, that is observed in the cortex at all levels, arises from mutual excitation within multiple populations of excitatory neurons. The governing point attractors are set by the mesoscopic states, which act as order parameters [31] that regulate the contributing neurons. The point attractors manifest a homogeneous field of white noise [12]. These properties show that the activity arises by widespread synaptic interactions, and that it is maintained and governed by a mesoscopic point attractor based on the recurrent excitatory collateral axons of cortical neurons [5, 9] in neocortex and in populations of excitatory neurons [12] in the olfactory system (Figure 1.9).

The most detailed study is focused on the periglomerular neurons in the outer layer of the olfactory bulb, which form a densely interconnected population receiving input from the receptors and giving output to the mitral cells. Each neuron excites thousands of others in its surround and receives from them along innumerable connections with distributed axonal distances and successive synaptic delays. The feedback is modeled by a rational approximation for a 1-dimensional diffusion process with the Laplacian operator s in the expression,

$$\exp[-(sT)^{0.5}], \tag{4}$$

where T is a lumped, distance-dependent time constant [12]. This accounts for the randomization of activity on each pass through the loop for each neuron. Owing to the sigmoid nonlinearity [10, 20] in the feedback, the population has two stable points, one at zero and the other at a nonzero level [6].

Thus, periglomerular pulse density is observed as noise at the microscopic level, but it is modeled as a d.c. bias that is stable under external perturbation at the mesoscopic level (top trace in Figure 1.10). The mechanism of stabilization of the point attractor by means of the sigmoid curve is shown in Figure 1.11, along with the amplitude histogram and power spectrum of periglomerular pulse density.

The bulb, nucleus and cortex interact by positive and negative feedback. The self-sustaining basal activity of the periglomerular neurons provides an excitatory bias to the mixed populations comprising the inner bulb, and also the cortices to which the bulb transmits. The three parts have characteristic frequencies in the gamma range (20-80 Hz) and feedback delays to each other as well as to the periglomerular neurons.

The inhibitory feedback between oscillators gives rise to negative or zero Lyapunov exponents. The excitatory feedback from the anterior nucleus to the periglomerular neurons gives a positive exponent. Aperiodic activity is sustained by the system in the absence of input. The basal activity again exhibits

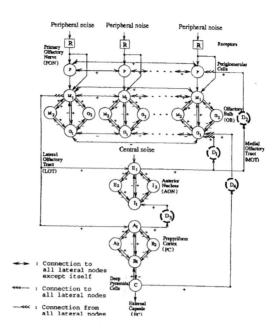

Figure 1.9.   Chaotic dynamics arises by feedback with delay among 3 coupled oscillators having incommensurate characteristic frequencies. The landscape is stabilized by additive noise. Each node is governed by equations (1) and (2). In digital models, noise serves to stabilize the chaotic attractors [25].

a Gaussian distribution of amplitude (Figure 1.12), but the operating rest point at zero wave density is on the rising phase of the gain curve, reflecting the inherent bistability of the KII set in strong contrast to the periglomerular KI set [20].

5.   Microscopic coding and noise

These observations in modeling the neuroanatomy and neurodynamics of olfactory cortex and simulating its spatiotemporal patterns of activity require inclusion of the distinction between high-dimensional "noise" at the microscopic unit level, low-dimensional chaotic "signal" at the mesoscopic EEG level [14], and the relations between them. The "noise" is essential for the maintenance of normal mesoscopic activity, for which it is a carrier, and the levels and spatiotemporal patterns of the "noise" are controlled by the mesoscopic activity acting as an order parameter [31] and as a "signal" [7, 17, 18, 25]. In turn, the mesoscopic activity of multiple mesoscopic domains organizes itself into macroscopic patterns that may occupy an entire cerebral hemisphere [19]. Microscopic neural coding is determined by recording pulse trains from single

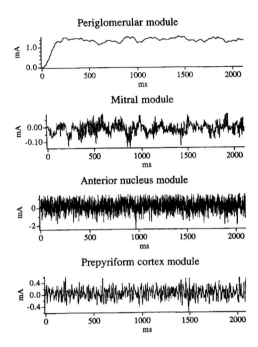

Figure 1.10. Simulations of output in the resting state by the KIII model are shown for activity patterns observed in 4 parts of the olfactory system in the rest state.

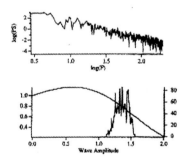

Figure 1.11. The stabilization of the KI set at the input layer of the olfactory bulb, PP, is on the right (upper) segment of the sigmoid curve, so that an increase in input amplitude leads to an increase in output amplitude, but the positive feedback gain is decreased, so the system returns to its nonzero rest level. That level is represented in a linear approximation by a real-valued root of zero, which corresponds to a pole at the origin of the complex plane. This pole governs the level of the background bias to the bulb and its targets of transmission.

cells during sensory stimulation and motor activation. The sensory code in all modalities is topographic from multiple receptor types distributed broadly and inhomogeneously over the receptor surface and extending their axons centrally as labeled lines. The intensity on each axon is conveyed by the pulse frequency.

Neural coding in cortical populations has been analyzed mainly from records of EEGs [13]. Corroborative evidence has been obtained by simultaneous recording of dendritic potentials and selected axonal potentials [12, 15]. The requirement has been imposed that (a) records be taken simultaneously from arrays of 16 to 64 electrodes placed on the cortical surface; (b) that the subjects be in the waking state and engaged in controlled sensorimotor behavior, such as performance of a conditioned reflex in response to conditioned stimuli; and (c) that brief epochs (on the order of 0.1 second) of EEGs be classified in respect to antecedent conditioned stimuli or consequent conditioned responses, solely on the basis of the spatiotemporal patterns of cortical activity that are extracted from sets of EEG traces [26, 28].

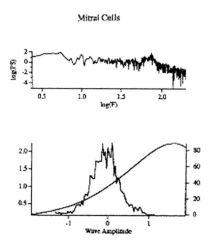

Figure 1.12.   Spectrum and amplitude histogram of olfactory bulb simulation. The gain curve is from equation 3, the derivative of the sigmoid function. The negative feedback loops of the three oscillators are stabilized on the left (lower) side of the sigmoid curve, where an increase in input leads to an increase in output and also to an increase in gain, giving the input-dependent destabilization. This mechanism for conditional stabilization and bistability is characteristic of the limit cycle attractor.

The content is distributed over the entire sensory area, so that every neuron is involved in every discrimination. Local areas with high amplitude EEG are no more or less important than those with low amplitude EEG [26, 27]. The fields have the forms of interference patterns resembling holograms, except that their mechanism of formation is nonlinear and unstable involving state transitions and there is no inverse by which to recreate the input from the created patterns. The mean firing rates have no value as state variables, because it is the set of relations of every local amplitude to all others that defines the content.

The conversion of a sensory stimulus in the microscopic code on input axons to a mesoscopic code distributed over the whole of the sensory cortex depends on the conjunction of three factors. One is the prior formation through learning of a nerve cell assembly of strengthened associational connections at the synapses of excitatory neurons onto other excitatory neurons [9, 12]. The second is a motivated state of the subject that is expressed in a steep sigmoid curve. The third is a surge of input that is widespread over the cortex and that includes input not only to the learned nerve cell assembly but also to many other neurons in the bulb, whose response has been attenuated by habituation to uninformative, distracting, and ambiguous input.

In these circumstances the destabilized cortical mechanism is guided by the assembly into the basin in a chaotic attractor landscape for the entire cortex, which is expressed by the spatial pattern of the output of the entire cortex.. That output is re-expressed in the microscopic form by the action potentials on the axons of the projection neurons that carry it to other areas of cortex or into the brainstem, where it has been shown to guide other interactive populations into basins of attraction formed in past learning experiences [4].

6. Chaotic attractor stabilization and classification enhancement by noise

New techniques for large-scale parameter optimization enable simulation of these time series, as well as their spatial patterns of amplitude modulation following learning, with the solutions of networks of coupled ordinary difference equations having a static sigmoid nonlinearity at the output of each node. The deterministic model is exquisitely sensitive to exceedingly small numerical changes in parameters, due to attractor crowding with basins of attraction shrinking to approximate the size of the digital numbers used for computation [35], and to lack of a shadowing trajectory [8, 30] due to the repeated destabilization and restabilization that requires translation of the real parts of the closed loop roots of the differential equations across the imaginary axis under piece-wise linearization [12, 20].

Considering that the olfactory system has both peripheral and central sources of noise, low- level Gaussian noise is added (Figure 1.9) at two significant points [7, 25]. This noise is designed to simulate the experimentally derived parameters of neurobiological noise: spatially independent and rectified at the input to simulate receptor input, and normally distributed with an excitatory bias but spatially coherent for the noise of central origin from other parts of the brain. The combined effect is to stabilize the model under input-induced state transitions and improves the simulations of EEGs and pulse densities from the olfactory system. Noise appears to do this by smoothing the landscape from digital representation in a nowhere differentiable manifold resembling rabbit fur to a smooth and well behaved landscape with a small number of large basins. These improvements are needed not only for EEG simulations but for optimizing the classification efficacy of the KIII model [32]. The Gaussian noise is simulated digitally with a random number generator, which supports robust and repeated access by repeated inputs to the learned basins of attraction in the simulated landscape through 1st order state transitions that are triggered by spatiotemporally patterned inputs given to the open nonautonomous KIII model, just as stimuli are received by brains that routinely function as open systems.

In using these data to model changes with associative learning a 2-D layer of 64 coupled oscillators is constructed to simulate the OB, and the simulated synaptic strengths between the coupled oscillators in the model are modified in accordance with modified Hebbian criteria [36]. The model also incorporates the biological property that the response energy during habituation to undesired input is not decreased, but is shifted from the gamma range to the theta range of cortical activity. Thereby the "background" noise is not filtered out but is used to enhance the effectiveness of the microscopic "foreground" signal by contributing to the destabilization simulating the respiratory wave (Figure 1.5). With education by examples under selective reinforcement, the spatial pattern of amplitude modulation of the common carrier changes from a previous pattern and stabilizes in a new pattern, whenever an example is given as a member of the learned class of input. The report given by the system generalizes to the class of membership, which conveys the meaning of a stimulus, and not its form that is unique to each instance of stimulus presentation.

Each new stimulus that the model is trained to discriminate gives rise to a distinctive new spatial pattern of output that consists of 64 scalar AM values of the carrier wave, so that each output is a vector for a point in 64-space. The degree of similarity or difference between patterns is expressed by a Euclidean distance in 64-space. Classification is based on clustering of the points derived from measurement of individual responses in the biological system [26]. The effectiveness of the classification depends equally on each of the available measurements from a topographic array of input channels, which shows that the information density is distributed and spatially uniform, not localized as with the point processes of sensory displays. These properties replicate the dynamics of the olfactory system operating in the chaotic mode [13, 36]. The most effective performance of the KIII model in pattern recognition is achieved when it is operating in a chaotic domain that has been stabilized by optimal levels of noise [32], and in which the noise is further optimized in the manner of stochasit resonance.

7.   Mesoscopic to macroscopic interface

This work is now in progress.

8.   Summary

The sensory input and motor output of brains is carried by action potentials at the microscopic level of brain function. The organization of the patterns of the 'units' that support behavior is done by large domains of the brain, as revealed by macroscopic brain imaging. The work of bridging between these levels is done at the intervening mesoscopic level by wave packets. These are spatially coherent domains of activity 1-2 cm in diameter, lasting 80-120 msec, and recurring at 2-7/sec. They are manifestations of local mean field activities of millions of neurons, which self-organize spatial patterns of activity in the form of amplitude modulation of a common aperiodic carrier wave. They show the intrinsic bistability of areas of cortex at the mesoscopic level, comparable to the bistability of the axon at the microscopic level. A sensory cortex receives input from sensory pathways during a receptive 'diastolic' period, then transits to a transmitting 'systolic' mode by a 1st order phase transition, in which state it combines the features of the sensory input in the context of past experience fixed in the synaptic matrix of the cortex by past learning and the current brain state determined by the neuromodulators in which the cortex is bathed. The cortical output is broadcast in the form of a distributed interference pattern, which is selected by the targets in much the way that radio receivers can be tuned to specific broadcast stations, unlike point-to-point transmissions in telephone networks. All this mesoscopic action is detected, measured, and understood through recording the EEGs generated by the neurons comprising and enslaved by the wave packets. The EEG is not in itself the agent of cohesion of the neural activity; it is the noise made by the millions of neurons that contribute to it. Any attempt to inject electric current from electrodes placed in cortex so as to simulate the function of a wave packet would be like trying to fly a plane by playing a tape recording of the plane taking off from a loudspeaker set up beside the plane parked on a runway. The upshot is that the design of new electronic devices can best be undertaken with the aim of realizing in hardware the properties of wave packets, not merely of action potentials.

Acknowledgments

This research was supported by grants from the National Institute of Mental Health MH 06686, and the Office of Naval Research N00014-90-J-4054.

References

[1] Abeles M (1991) Corticonics: Neural Circuits of the Cerebral Cortex. Cambridge UK: Cambridge University Press.

[2] Aertsen A, Erb M, Palm G (1994) Dynamics of functional coupling in the cerebral cortex. Physica D 75: 103-128.

[3] Barrie JM, Freeman WJ, Lenhart M (1996) Modulation by discriminative training of spatial patterns of gamma EEG amplitude and phase in neocortex of rabbits. Journal of Neurophysiology, in press (July).

[4] Bressler SL (1988) Changes in electrical activity of rabbit olfactory bulb and cortex to conditioned odor stimulation. Journal of Neurophysiology 102: 740-747.

[5] Chang H-J, Freeman WJ (1996) Parameter optimization in models of the olfactory system. Neural Networks 9: 1-14.

[6] Chang H-J, Freeman WJ (1998a) Optimization of olfactory model in software to give 1/f power spectra reveals numerical instabilities in solutions governed by aperiodic (chaotic) attractors. Neural Networks 11: 449-466.

[7] Chang H-J, Freeman WJ (1998b) Biologically modeled noise stabilizing neurodynamics for pattern recognition. International Journal of Bifurcation and Chaos 8: 321-345.

[8] Dawson S, Grebogi C, Sauer T, Yorke JA (1994) Obstructions to shadowing when a Lyapunov exponent fluctuates about zero. Physical Review Letters 73: 1927-1930.

[9] Douglas RJ, Koch C, Mahowald M, Martin KAC, Suarez HH (1995) Recurrent excitation in neocortical circuits. Science 269: 981-985.

[10] Eeckman FH, Freeman WJ (1991) Asymmetric sigmoid nonlinearity in the rat olfactory system. Brain Research 557: 13-21.

[11] Eisenberg, J., Freeman, W.J. and Burke, B. Hardware architecture of a neural network model simulating pattern recognition by the olfactory bulb. Neural Networks 2: 315-325, 1989.

[12] Freeman WJ (1975) Mass Action in the Nervous System. New York: Academic

[13] Freeman WJ (1987) Techniques used in the search for the physiological basis of the EEG. In: Gevins A, Remond A (eds) Handbook of EEG and clinical Neurophysiology Vol 3A, Part 2, Ch. 18. Amsterdam: Elsevier.

[14] Freeman WJ (1988) Strange attractors govern mammalian brain dynamics, shown by trajectories of electroencephalographic (EEG) potential. IEEE Trans. Circuits & Systems 35: 781-783.

[15] Freeman, WJ (1992) Tutorial in Neurobiology: From Single Neurons to Brain Chaos. International Journal of Bifurcation and Chaos 2: 451-482.

[16] Freeman WJ (1995) Societies of Brains. A Study in the Neuroscience of Love and Hate. Hillsdale NJ: Lawrence Erlbaum Associates.

[17] Freeman WJ (1996) Random activity at the microscopic neural level in cortex ("noise") sustains and is regulated by low-dimensional dynamics of macroscopic cortical activity ("chaos"). International Journal of Neural Systems 7: 473-480.

[18] Freeman WJ (1998) The regulation and use of microscopic neural noise by macroscopic chaos within populations of neurons in brains. pp. 89-105. Proceedings, 19th Nihon University International Symposium "Order and Non-Order". Ito Y, Kawakami I, Konno K, Matunaga Y, Shimada I, Tsubokawa T (eds.) Singapore: World Scientific.

[19] Freeman WJ (1999) How Brains Make Up Their Minds. London UK: Weidenfeld & Nicolson.

[20] Freeman WJ (2000) Neurodynamics. An Exploration of Mesoscopic Brain Dynamics. London UK: Springer.

[21] Freeman W J, Baird B (1987) Relation of olfactory EEG to behavior: Spatial analysis: Behavioral Neuroscience 101:393-408.

[22] Freeman WJ, Barrie JM (1994) Chaotic oscillations and the genesis of meaning in cerebral cortex. In: Buzsaki G, Llin s R, Singer W, Berthoz A, Christen Y (eds.) Temporal Coding in the Brain". Berlin, Springer-Verlag, pp 13-37.

[23] Freeman WJ, Barrie JM (2000) Analysis of spatial patterns of phase in neocortical gamma EEGs in rabbit. Journal of Neurophysiology 84: 1266-1278.

[24] Freeman WJ, Barrie JM, Lenhart M, Tang RX (1995) Spatial phase gradients in neocortical EEGs give modal diameter of "binding" domains in perception. Abstracts, Society for Neuroscience 21: 1649 (648.13).

[25] Freeman WJ, Chang H-J, Burke BC, Rose PA, Badler J (1997) Taming chaos: Stabilization of aperiodic attractors by noise. IEEE Transactions on Circuits and Systems 44: 989-996. .

[26] Freeman WJ, Grajski KA (1987) Relation of olfactory EEG to behavior: Factor analysis: Behavioral Neuroscience, 101: 766-777.

[27] Freeman WJ, Van Dijk B (1987) Spatial patterns of visual cortical fast EEG during conditioned reflex in a rhesus monkey. Brain Research: 422: 267-276.

[28] Freeman WJ, Viana Di Prisco G (1986) Relation of olfactory EEG to behavior: Time series analysis. Behavioral Neuroscience 100:753-763.

[29] Freeman WJ, Yao Y, Burke B (1988) Central pattern generating and recognizing in olfactory bulb: A correlation learning rule. Neural Networks 1: 277-288.

[30] Grebogi C, Hammel SM, Yorke JA, Sauer T (1990) Shadowing of physical trajectories in chaotic dynamics: Containment and refinement. Physical Review Letters 65: 1527-1530.

[31] Haken H (1983) Synergetics: An Introduction. Berlin: Springer-Verlag.

[32] Kozma R, Freeman WJ (2001) Chaotic Resonance: Methods and applications for robust classification of noisy and variable patterns. International Journal of Bifurcation and Chaos, in press, June 2001.

[33] Singer W, Gray CM (1995) Visual feature integration and the temporal correlation hypothesis. Annual Review of Neuroscience 18: 555-586.

[34] Softkey WR, Koch C (1993) The highly irregular firing of cortical cells is inconsistent with temporal integration of random EPSPs. Journal of Neuroscience 13: 334-350.

[35] Tsang KY and Wiesenfeld K (1990) Attractor crowding in Josephson junction arrays. Applied Physics Letters 56: 495-496.

[36] Yao Y, Freeman WJ (1990) Model of biological pattern recognition with spatially chaotic dynamics. Neural Networks 3: 153-170.

Chapter 2

COMPUTATIONAL AND INTERPRETIVE GENOMICS

Steven A. Benner
Department of Chemistry,
University of Florida,
Gainesville FL 32611

As one of its goals, this conference seeks to build bridges between two scientific cultures, represented on one side by computer scientists and physical chemists, and on the other by biologists and organic chemists. These two group are both interested in how computation can help us better understand biological systems. The motivation for joining the two cultures is heightened by the growth of genomic sequence databases, which add greatly to the amount of information describing the structure of biological macromolecules, in particular, DNA and proteins.

This lecture presents one particular view of the nature of the separation that makes these bridges difficult to build. As the first step, we must recognize that the two cultures, although they often use the same words, do not necessarily speak the same language [1]. This means that even when those from the two cultures agree on the problems that need to be solved, they might disagree on the approach to the solution, the forms that solutions might take, or even the criterion that might be used to identify an acceptable solution. This means, of course, that the two cultures often do not interact productively.

As this conference unfolded, it became clear that it itself illustrated this difficulty. As a result, this talk was revised and presented, later in the conference, to address the issue explicitly. By doing so, we hoped to help bridge the two cultures by illustrating the contrast between the biological/chemical approach and the computational/physical approach to a problem well known both cultures, predicting the three dimensional conformation (or "fold") of a protein from sequence data.

P.M. Pardalos and J. Principe (eds.), Biocomputing, 25-43.
© 2002 *Kluwer Academic Publishers. Printed in the Netherlands.*

The protein structure prediction problem

Protein structure prediction provides an archetypal example of a problem that has been addressed separately by computationally-oriented scientists and biologically-oriented scientists. Briefly stated, the problem relates to the relationship between the *constitution* of an organic molecule, a chemical formula stating which atoms are bonded to which in the molecular structure, and the *conformation* of an organic molecule, a statement about how atoms are disposed in three dimensional space. The conformation of a protein is sometimes called the *fold*, as the linear chain of amino acids in a protein often folds in its native state to give a compact, or *globular* structure.

Many groups have sought to predict the conformations (folds) of proteins from their constitutions (sequence). The effort reflects, in part, the recognition that conformation is more closely related to *behavior* than is the sequence itself, where behavior is defined as an observable property of the molecule (a molecular phenotype). Behavior, in turn, is interesting because it is more closely related to *function* than either fold or sequence. Function, of course, is what truly interests biologists, and is defined as those behaviors that contribute to the fitness of an organism. Under Darwinian theory, natural selection acting on random variation is the only mechanism for obtaining functional behavior in a biological molecule.

As technology to sequence genes became more powerful, it became clear that protein sequences would become available far more rapidly than information about protein behavior and function. The hope was therefore to predict conformation from sequence, and therefore take a large step towards inferring something about function from sequence, at least at the level of hypothesis. Sequence-derived hypotheses would then, it was hoped, direct experimental work more efficiently than was possible without a predicted conformation. Hence the motivation to predict conformation from sequence.

The frontal assault on protein structure prediction

As early as the 1960's, the protein folding problem was perceived to be a good area to apply the physical science paradigm to biology. The early heroes in the field, including Scheraga [2], Karplus, and others [3], resurrected a much older idea introduced by Westheimer and Mayer in the 1940's to apply molecular mechanics approaches to address the problem of conformation [4]. Briefly, the strategy was to build a computational representation of the protein molecule of interest. To this would be added a model for how atoms in the representation might attract and repel each other (a "force field"). The task was then to simulate the motion of the protein in a "molecular mechanics" experiment, where Newtonian-like movement of the protein was permitted. Given sufficient simulation time, a computer would "search" the hypersurface relating potential energy to conformation in the molecular representation, and (it was hoped) find a global energy minimum that would reflect the natural conformation of the protein.

In the original Westheimer implementation of molecular mechanics, computations were done by hand-cranked calculator. Increasingly powerful computers were later directed towards this "frontal assault" on the protein structure prediction problem. The process is continuing. For example, IBM recently announced the dedication of a supercomputer (Blue Gene) specifically to this assault [5].

To date, the assault has failed. We understand, in an incomplete way, why this is so, and to some extent why is must have been so. At the very least, physical theory requires the calculation to find the conformation of the protein having the lowest *free* energy, not the lowest *potential* energy. The theory therefore requires that entropy be considered. Proteins fold in water (or still worse, at a water-membrane interface), and water is a strongly interacting solvent. Indeed, changes in entropy associated with changes in conformation of solutes dissolved in water are often dominated by changes in the entropy of the water that solvates those solutes. Certain to obstruct the frontal assault is the fact that there is (to date) no good model to describe the free energy of interaction between solutes and water, enthalpy or entropy.

Even if water were not a problem, there were other reasons why physics expected the frontal assault to fail. Physical theory specifies the level of resolution that is required for a computational model of any molecule (including a protein) to have predictive and manipulative value. Most optimistically interpreted, theory insists that the atom must be the unit of interaction. Less optimistically, interaction between sub-atomic units (electrons, for example) might be important. But for certain, physical theory does not expect that models with poorer-than-atomic resolution will have predictive and manipulative value. This means that *every* atom of the protein (and a typical protein contains perhaps 2000 of them) must be incorporated into the model for it to be likely (under physical theory) to be suitable to support the prediction of conformation. Indeed, physical theory would seem to require that atoms from the solvent (water) be incorporated as well.

This generates a computational problem of extraordinary complexity. Even the best computers would require time far in excess of the lifetime of the galaxy to compute the energy of an individual state of a protein, if every interaction must be considered. At best, computational models approximated this energy by (most commonly) considering only spatially proximal pairs of atoms. Even this is a tremendous computational task if all atoms (including water) are considered, tens of orders of magnitude beyond the capabilities of computers of the past. And even if the approximations were adequate, well known problems with molecular mechanics simulations (such as the propensity of such simulations to become trapped in local energy minima) frustrated many efforts to obtain the global energy minimum.

But the hardness of the computational task was still not the most serious problem. Physical theory requires that the nature of microscopic interactions must be understood before a simulation could begin. And the nature of the microscopic interactions between atoms was not known, except as an approximation. This remains true even today.

Over the past forty years, various tests have been applied to benchmark the ability of molecular theory to accurately describe the conformation of molecules much smaller than proteins, and in solvents far less interactive than water. In constructing these tests, parameterization has been permitted, often quite liberally. For example, in substituted cyclohexanes in weakly interacting solvents, measurements derived from experiments were used to parameterize the force field, which was then applied to predict the conformational behavior of new substituted cyclohexanes [6]. Over a decade or so, the force fields became sufficiently well parameterized to predict the conformational behavior of substituted cyclohexanes reasonably well, at least in weakly interacting solvents.

These tests provide a view of the level of parameterization that is likely to successfully move from constitution to conformation in organic molecules in general, and proteins in particular. Proteins are, of course, far more challenging than substituted cyclohexanes, which have exactly one ring and no opportunity to unfold. Many other conformational challenges, more complicated than substituted cyclohexanes but simpler than proteins, remain unmet. One particularly relevant case relates to the computational modeling of the packing of organic crystals [7, 8, 9], a problem directly analogous to (but still simpler than) the packing of the hydrophobic cores of protein molecules. Another unsolved problem challenges the modeller to predict the solubility of an amino acid in water, again directly relevant to the protein folding problem.

Considering what we know about the problem, it would have required something magical (that is, aphysical) had it been possible to use molecular mechanics to predict protein conformation. There might indeed be a parameterized model that permits a calculation of a potential energy using an approximate model for solvent acting on a model of a protein specifying (for example) only the backbone atoms, or a model projected upon a one nanometer lattice, in a way that is predictive of the conformation of a real protein having the lowest free energy in water. But all that we knew about chemistry suggested that this was unlikely, indeed, very unlikely.

Evaluating tools for the frontal assault

It is remarkable how little of this discussion is reflected in the literature generated by computational scientists mounting a frontal assault on the protein folding problem. For example, IBM's Blue Gene project seems to be based on the notion that increasing the speed of "number crunching" by a factor of forty thousand will make a dramatic improvement in the ability of a force-field based model to predict the folded conformation of a protein [5]. This improvement is not even close to what is needed.

Another example comes from an ongoing project known by the title "Critical Assessment of Structure Prediction" (CASP) [10]. The CASP project challenges computational scientists to build a three dimensional model for proteins whose folds are known, but not released before the models are built. The CASP project is designed and evaluated by scientists drawn from the computational culture.

This is reflected in particular in the way that the predicted models for protein fold are evaluated. Most prominent among the evaluation method is to determine the extent to which placement of the backbone atoms in the model differs, root mean square, from the placement of the backbone atoms in an experimentally determined structure [11]. This makes perfect sense to someone from the computational science culture. But to someone whose grounding is in organic and biological chemistry, this is akin to asking for the construction of the Empire State Building by someone who has not yet learned the strength of a steel beam. It would make far more sense to have such a contest to predict the packing of crystals of small organic molecules, or the solubility of small organic molecules in water.

Getting perspective: Why are we doing this anyway?

Approximate solutions to any problem may be useful, of course. Here, the acceptable level of approximation depends on the purpose to which the solution will be applied. In implementing their frontal assault, it has been easy to lose track of this question: *Why* do we want to predict the conformation of a protein from sequence data?

In the literature, the logical progression that "sequence-gives-fold-gives-behavior-gives-function" is invoked. It makes a coherent explanation, unless one recognizes that one piece of the logic is missing. Specifically, even if we know the conformation of a protein, it turns out that we still cannot generate a statement about behavior or function. Chemical theory is inadequate to make a statement *de novo* about any particular property of any protein in solution, even if given a three dimensional model of the conformation of a protein at any arbitrarily high level of resolution. We cannot say what the protein binds to. We cannot say whether it catalyzes a reaction. If we are told that it does, we cannot say which reaction it catalyzes, or how fast, or in what way. Structure, even after all of its components (including conformation) are known, does not inform us about the behavior of a molecule, meaning that we cannot infer function from structure either.

Homology: The "why" for protein structure prediction

So why *do* we want to know the conformation of a protein? There are many reasons, even in the absence of theory that allows us to convert conformational information directly to behavior. The one most closely related to function concerns the desire to use conformation to identify *homologous* proteins, two (or more) proteins related to each other via common ancestry. Brothers and sisters are close homologs, in terms of human relationship. So are humans and amoebae, judging from a wealth of molecular evidence. In between come chimpanzees, rhesus monkeys, cattle, kangaroos, frogs, fish, and fruit flies.

What is the relationship between a search for homology and predicting the folds of proteins? The answer is simple, and is based on an empirical observation. Homologous proteins are generally observed to have analogous folds [12]. The generalization is striking. If the structures of two homologous proteins are superimposed, the root mean squared deviation in the position of two backbones is generally less than 0.2 nm, even after the sequences of the proteins have changed so much that their similarities can scarcely be recognized (at the level of sequence) [13]. This leads to the converse notion. Two proteins having analogous folds are much more likely to be homologous than two proteins not having analogous folds.

Why are biologists interested in identifying homology between proteins? In part, this is because the evolutionary history of a protein family contains information about function. Two proteins related by common ancestry need not contribute to fitness in the same way (the Darwinian definition of "function"). But they very often bind to analogous substrates, or catalyze analogous reactions, or do other things in analogous ways. Knowing the fold of protein i can frequently allow a biologist to decide that it is homologous to protein j, and from this decision, infer that certain behaviors of protein i are analogous to certain behaviors of protein j. This might provide the key to an understanding of the function of protein i, especially when something is known about the function of protein j.

So here is a reason to predict fold from sequence (to learn about homology), a reason that is logically coherent from start to finish. Predicted folds could be, in principle, powerful tool to confirm or deny the possibility that two proteins are related by common ancestry. This reason is understood by most biologists, but by few computational scientists.

This reason also allows us to say when the prediction problem is "solved", at least for this particular technological purpose. The problem is solved, for this purpose, when the predictive tool can confirm or deny the possibility that two proteins are related by common ancestry. To the biologist (at least one that wishes to understand homology), a good standard for judging a prediction "contest" (for example) is: Does the predicted fold allow me to confirm (if appropriate) or deny (if appropriate) homology? Evaluating a prediction for its ability to do this makes far better sense than evaluating a prediction based on its root mean square fit to a reference structure.

As it turns out, homology can be predicted or denied with rather low resolution models. One needs to know the placement of standard secondary structural elements (helices and strands) in a protein sequence, and some idea of how they are packed. If the placement of helices and strands is (reasonably) accurate, and the tertiary packing of these is (reasonably) accurate, then inferences about homology can be drawn. And the advantage of such *de novo* predictions over alternative methods that require a target structure (e.g., threading) is that they can both confirm *or* deny the possibility of distant homology.

As the papers in this conference suggested, very few computational chemists know that tools that predict the structure of proteins at this level are available today. Indeed, they have been tested and shown to be useful for more than a decade.

Going around the very hard computational problems to predict protein fold

To understand how to generate useful models of folded structure from protein sequence data, we must turn to evolutionary history. This, it turns out, suggests an approach to the protein structure prediction problem entirely different from the "frontal assault" engineered by the physical scientist.

As noted above, the process of divergent evolution creates families of proteins descendant from common ancestors. As proteins divergently evolve from those ancestors, natural selection requires them to remain "fit". Indeed, the only way for the sequence of a protein to enter the database (and for us to know about it) is if the organism that contained it survived, selected a mate, and reproduced. The principal prerequisite for fitness in a protein is a fold. Therefore, proteins diverging from a common ancestor generally conserve their folds.

This means that during the evolution of protein sequences, amino acid replacements are not "fixed" in the population as they would be if proteins were formless, functionless organic molecules. Amino acids important to the fold suffer substitution differently from those that are not. A signal should lie in the pattern of protein sequence divergence that should tell us something about the protein's form and its function. And this is how the *ab initio* protein structure prediction problem has been "solved", at least to a level that permits the predictions to be useful.

Let us amplify on this solution using the language of the mathematical biologist. Virtually all treatments of protein sequence evolution have been based on a "first order stochastic" model [14]. This model (generally) assumes:

- Future mutations are independent of past mutations

- Mutations occur independently at different positions

- The probability of substitution reflects a 20 x 20 "log odds" matrix characteristic of the database as a whole

- Gaps can be scored using a penalty + increment formula

This model is convenient. In particular, it fits well the dynamic programming tools developed by Needleman, Wunsch [15], Smith, Waterman [16], and others [17], tools that have been quite useful in identifying optimal alignments of two protein sequences (given a particular theory of evolution).

We know from empirical studies that such models only poorly approximate reality, however. In reality, future mutations in folded proteins do not accumulate independently of past mutations. Mutations at different positions are correlated [18], often because amino acids distant in the sequence are in contact in the three dimensional fold. Gaps display a Zipfian (not an exponential) length distribution [19]. Individual residues accept amino acid replacements depending on functional constraints specific to that residue.

The fact that commonly used stochastic models only poorly approximate how proteins actually evolve need not be paralyzing. Rather, the stochastic models can be used as a description of how a protein *would* evolve if it were a formless, functionless string of letters. The difference between how proteins actually evolve and how stochastic models treat protein evolution is a signal that contains information about fold and function.

The Exhaustive Matching

Computer tools were necessary to learn how to extract a signal about fold and function from the difference between how proteins actually replace amino acids during divergent evolution under functional constraints, and how simple mathematical models describe that evolution. To develop the first of these, we were joined with Prof. Gaston Gonnet, first at Waterloo and later at the Swiss Federal Institute of Technology, to develop computer software tools designed to empower the biologist to use genome sequence data. The most useful of these tools was the bioinformatics workbench known as DARWIN [20]. DARWIN provides a high level programming language, a set of analytical tools (many derived from the symbolic computation program known as Maple, which Prof. Gonnet had helped develop), and a database management environment that empowered biologist and biochemists, enabling them to ask and answer questions about protein sequences.

In 1992, well before the "age of the genome" began, DARWIN allowed us to generate a comprehensive empirical test of a "first order stochastic" model of protein evolution [21]. Many of the features of the results of this test have been reviewed and discussed in the literature [14]. Let us just consider a few of them, beginning with the fact that positions in a protein do not suffer replacement independently. In fact, there is a strong correlation (for example) between substitution at position i and position $i+1$. Specifically, if the residue at position i is conserved, then the residue at position $i+1$ is generally more likely to be conserved as well. This is well understood as a consequence of the fact that positions on the surface of a protein fold are more likely to be mutable than residues buried within the fold, and that a position on the surface of the fold are more likely to be next to another position on the surface than is the average position.

Remarkably, however, the exhaustive matching found general exceptions to this rule, exceptions that are embedded in the intrinsic relationship between amino acid structure and the fold. Specifically, if position i contains a conserved proline (P) or glycine (G), it was generally the case that the adjacent position would have suffered more variation than the average position. What was the physical chemical basis behind this higher order behavior? A structural hypothesis was possible. A conserved Pro or Gly often defines a critical turn. Turns normally occur on the surface. Thus, a conserved Pro or Gly is likely to be adjacent to a mutable surface residue.

This hypothesis could be readily transformed into a rule to predict conformation. If position i contains a conserved Pro or Gly and $i+1$ or $i-1$ has suffered a replacement, then the sequence turns at this point. This proved to be one of many ways in which the behavior of homologous protein sequence undergoing divergence under functional constraints, different from that predicted by a simple stochastic model, holds information about conformation. More generally, it has now become clear that a detailed, empirically-based understanding of divergent evolution helps develop tools for extracting secondary structural information from a set of aligned homologous sequences.

These types of rules underlie "evolution-base structure prediction" (ESP) methods. Intriguingly, they do not solve any of the difficult problems encountered in the "frontal assault". We need not obtain a force field, or search a high dimensional energy surface to find a global energy minimum, to implement an ESP prediction. All that we must do is extract information from patterns of variation and conservation between the sequences of homologous proteins diverging under functional constraints, and understand those patterns in the light of a fundamental understanding of how amino acid structure influences polypeptide conformation. Surprisingly, ESP tools work rather well, as has been confirmed by various CASP competitions, as well as in ca. two dozen *ab initio* predictions which have been used to confirm or deny homology. The popular PHD program implements these tools, albeit in a non-transparent way (as a neural network) [22].

Starting in 1990, we and others published a series of *ab initio* predicted structures for protein families. These are reviewed in reference 14, and include proteins important for biomedical research, including protein kinase, protein tyrosine phosphatase, protein serine phosphatase, hemorrhagic metalloprotease, the Src homology 2 and 3 domains, the pleckstrin homlogy domain, phospho-beta-galactosidase, synaptotagmin, ribonucleotide reductase, and heat shock protein 70. In each case, an *ab initio* prediction was published before any structure was known for any member of the family of proteins. The *ab initio* prediction was used to confirm/deny homology with other families of proteins. For many, the prediction then created insight about function.

Non first-order-stochastic behavior has been useful throughout these prediction exercises. For example, we needed to assemble the beta sheets predicted in the fold of protein kinase. Protein kinase was known to have a sequence "motif" found in many other kinases built on the parallel beta sheet of the "Rossman fold".

Gly-Xxx-Gly-Xxx-Xxx-Gly-(Xxx-Xxx-Xxx-Xxx-Xxx) _Lys

Many groups assumed that the presence of this motif meant that protein kinase and adenylate kinase were homologs. Following the empirical generalization discussed above, by 1990, five groups had modeled protein kinase to have a fold analogous to the fold of adenylate kinase [23]. These models all assumed that a parallel beta sheet lay at the core of the folded protein kinase, just as it does in adenylate kinase.

Higher order analysis of sequence divergence, seeking the signal in the difference between real evolution and mathematical models for it, prediction denied distant homology [24]. The motif was predicted to lie between two antiparallel strands in protein kinase (whereas it was known to lie between a strand and a helix in a parallel sheet in adenylate kinase) (Figure 2.1). The antiparallel orientation of the two strands was predicted in protein kinase based on charge compensatory covariation at positions 87 and 108. At position 108, a negatively charged amino acid was replaced for a neutral amino acid at the same time as a positively charged amino acid was replaced for a neutral amino acid at position 87. Evidently, positions 87 and 108 are "talking to each other" during divergent evolution. We predicted that their side chains would be near in space; it followed that the predicted strands must be antiparallel.

It is useful to compare this result with results that might be obtained with threading, homology modeling, or BLAST searches, some of the tools used in

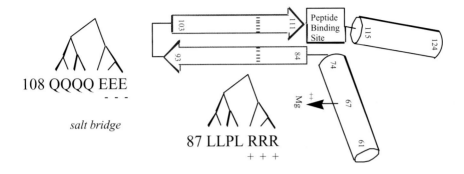

Figure 2.1. Positions 87 and 108 undergo correlated change during the divergent evolution of protein kinase (inset evolutionary trees), leading to a prediction that the two lie near in space in the folded structure of protein kinase. This prediction turned out to be correct, and allowed the correct assembly of the antiparallel sheet in the first domain of the folded protein.

conventional structure prediction/homology testing in modern computational biochemistry. Such approaches (at best) can say that two protein *might be* homologous. They can never exclude the possibility of distant homology. Not surprisingly for a tool that can say "yes" (but never "no"), threading, homology modeling, or BLAST searches frequently embed a few true "hits" among a large number of false positives. This makes them deficient.

ESP tools are now widely used to make predictions. These have been used to detect distant homologs (or to deny distant homology). Further, they have been used frequently to make statements about function, coupling a structure prediction to a statement about homology to a statement about analogy in behavior. Some examples are listed in Table 2.1.

To practicing biotechnologists, for at least some of the applications that they have in mind, ESP methods solve the protein structure prediction problem. But this is a biologist's solution to the problem, not a computational scientist's solution. Many computational scientists find the form of the solution unsatisfactory. After all, ESP methods do not readily give structure predicted to atomic resolution. They do not use force fields. They do not search high dimensional energy surfaces. To many computational chemists, ESP *cannot* be a solution, simply because its form is not the expected one.

This cultural clash was evident in the very first CASP project. Only two *ab initio* targets were presented. The ESP approach was used to predict both. Both predictions were valuable for the purpose outlined above, as acknowledged by the judges of the *ab initio* part of the project [25]. Yet some computational biologists still insist that CASP 1 yielded no correct structure prediction for the

Table 2.1. Some Useful Predictions using Evolution-based Structure Prediction (ESP)

Before CASP

- Protein kinase is not a homolog of adenylate kinase [24]

- Hemorrhagic metalloproteinase is a homolog of collagenases [14]

CASP 1:

- Phospho-beta-galactosidase folds like triosephosphate isomerase [30]

- Synaptotagmin forms a 7-strand antiparallel sheet [30]

CASP 2:

- Heat shock protein is a homolog of gyrase, and binds ATP [31]

And beyond:

- Ribonucleotide reductase that uses vitamin B12 is a homolog of the ribonucleotide reductase that uses iron [32]

- Calcineurin uses zinc as a cofactor [33]

- Leptin is a cytokine homolog, but not an obesity gene in humans [34]

two *ab initio* targets. The problem? The method used to generate the correct structural models *ab initio* did not have the form that they expected.

From structure to function

The ESP technology has opened up a new range of application of computational biology, one that now grasps function as well as structure. The difference between how proteins divergently evolve under functional constraints and how simple mathematical tools model that divergences tells us a lot more about biological function than simply a fold.

Extracting this information involves improving the mathematical model that describes protein sequence divergence. The first level of tools continues the tradition of empowering the biologist. Together with scientists at EraGen Biosciences, we have developed the Master Catalog, a naturally organized genome sequence database. The Master Catalog is based on the fact that after all of the genomes of all of the organisms on Earth are sequenced, all of the proteins will be organized into fewer than 10^5 nuclear families, joined by bridges indicating distant homology to give perhaps as few as 10^3 superfamilies of proteins in all.

The Master Catalog is a naturally organized database that contains an evolutionary model for each of these nuclear protein families. It builds a preconstructed multiple sequence alignment and evolutionary tree for each family. It then reconstructs sequences of ancestral proteins throughout the tree. These ancestral proteins were the intermediates in the divergent evolution of the protein family as it is now represented by modern sequences, placed at the "leaves" of the evolutionary tree.

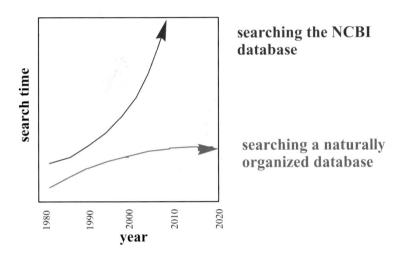

Figure 2.2.   The naturally organized database embodied in the Master Catalog makes searching more rapid than BLAST.

The natural organization of the Master Catalog provides it with its first advantage: It is very easy to update and search (Figure 2.2). New sequences can be added to the existing modular families, wedged into pre-existing multiple sequence alignments. Further, a "founder sequence" near the root of the tree, is a surrogate for all of its descendants. One need not search an exponentially growing number of sequence entries to find a homolog. Rather, one simply need to search an asymptotically growing number of founder sequences. Over 50% of these have almost certainly been found already. This means that the search of a naturally organized database containing all of the sequences of all of the organisms on Earth will be less than twice as expensive than the search of the incomplete database available to us today.

MASTER CATALOG and functional bioinformatics

The Master Catalog has other advantages. First, because the evolutionary models are precomputed, the biologist starts by asking biological questions, not

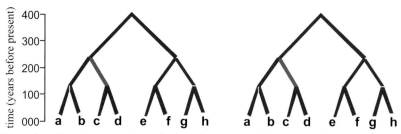

Two families in organism show adaptive evolution at the same time.
Hypothesis: The proteins interact as they function.

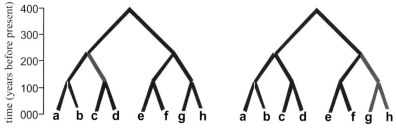

Two families do not show adaptively evolution at the same time.
Hypothesis: The proteins do not interact as they function.

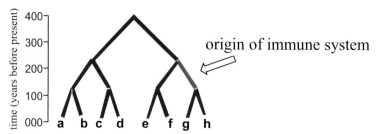

Duplication/adaptive evolution concurrent with new emergent physiolo
Hypothesis: The protein is involved in the immune system

Figure 2.3.   The naturally organized database embodied in the Master Catalog allows function to be inferred from protein sequence data by correlating events in the molecular record with events in natural history.

by cutting and pasting. Further, just as correlated change in a protein indicates what amino acids are in contact, correlated change in more than one protein family indicate what proteins are in contact when they function. This provides a new method to determine metabolic and regulatory pathways from sequence data (Figure 2.3).

One of many applications of this approach came when it was asked whether leptin should be the target of a drug discovery effort to treat obesity. Leptin

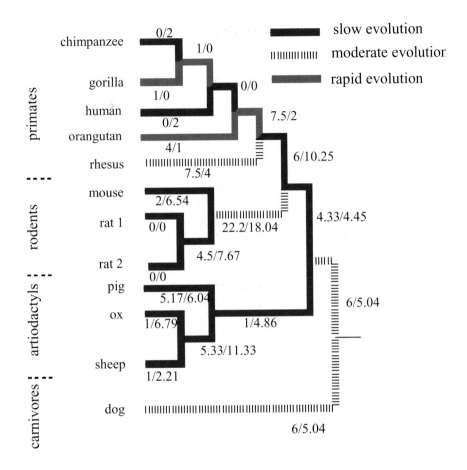

Figure 2.4. The naturally organized database embodied in the Master Catalog allows the user to identify examples where function may have changed in natural history. Here, primate leptin is inferred not to have the same function as mouse leptin, an inference with substantial consequences to pharmaceutical firms that may have chosen leptin as a target to treat human obesity.

was discovered in mice using genetic experiments; mice lacking leptin ate too much and became fat. The human ortholog of leptin was rapidly cloned from human tissue after the sequence of mouse leptin became known. Amgen paid perhaps $20 million for the rights to human leptin as a pharmaceutical target, and many hundreds of millions of dollars were invested in leptin by pharmaceutical companies around the world in the hope of developing a drug that would treat human obesity [26].

Structure prediction suggested that leptin is a cytokine, a protein related to interferon and interleukin. But an evolutionary analysis showed that since the divergence of primate leptin from a protein that was encoded by the genome of the most recent common ancestor of mice and men, leptin underwent an episode of rapid sequence evolution (Figure 2.4).

Rapid change in sequence implies rapid change in behavior. This, in turn, implies rapid change in function. Based on this evolutionary analysis, we predicted in 1997 that leptin in humans does not have the same physiological function in man as in mouse [26]. While this prediction remains non-confirmed, papers are now emerging with titles such as "Whatever happened to leptin?" [27], noting that "the hormone's precise physical role seems to vary from species to species." Indeed.

Analogous statements can be made about other pairs of orthologs from mammalian species [28, 29]. Over the next few decades, the tools that began with protein conformation prediction will exploit natural history to understand biological function from the molecule to the ecosystem.

Summary

From protein structure prediction to interpretive genomics, we have outlined an approach to doing science that is quite different from that appreciated by computational scientists. Computational scientists appreciate algorithms subject to formal proof. Failing that, they insist on statistical metrics and automatic heuristics. They can be uncomfortable with the notion that a useful hypothesis need not lend itself to direct mathematical test, or that success in technology might have a non-computational form.

None of these are problems, unless we allow them to be so. Obviously, the diversity represented by biologists on one side and computational scientists on the other can be complementary, and therefore very product. We hope that this lecture has helped scientists from both sides understand that those from the other side are not "wrong", but just "different", and to recognize the value in this difference.

References

[1] Kuhn, T. S.,*The Structure of Scientific Revolutions*, Chicago University Press, Chicago. 1962.

[2] Scheraga, H. A., Structural studies of ribonuclease III. A model for the secondary and tertiary structure. *J. Am. Chem. Soc.* 1960; 82, 3847-3852.

[3] Fasman, G. editor, *Prediction of Protein Structure and the Principles of Protein Conformation*, Plenum: New York,1989.

[4] Westheimer, F. H., Mayer, J. E., The theory of the racemization of optically active derivatives of diphenyl. *J. Chem. Phys.*, 1946; 14, 733-738.

[5] Bush, S., Meaning of life. *Electronics Weekly*, 07/12/00.

[6] Echeverria, G. A., Baron, M., Punte, G., Ab initio and in-crystal geometry of trans-1,4-dibromo-1,4-dicarboxymethylcyclohexane. *Structural Chem.* 2000; 11, 35-40.

[7] Dunitz, J. D., Filippini, G., Gavezzotti, A.,Molecular shape and crystal packing: A study of $C_{12}H_{12}$ isomers, real and imaginary. *Helv Chim Acta* 2000; 83, 2317-2335.

[8] Dunitz, J. D., Filippini, G., Gavezzotti, A., A statistical study of density and packing variations among crystalline isomers. *Tetrahedron* 2000; 56, 6595-6601.

[9] Lommerse, J. P. M., Motherwell, W. D. S., Ammon, H.L., et al., A test of crystal structure prediction of small organic molecules. *Acta Crystallogr B* 2000; 56, 697-714 Part 4.

[10] Moult, J., Hubbard, T., Fidelis, K., Pederson, J. T., Critical assessment of methods of protein structure prediction (CASP). Round III. *Proteins Struct. Funct. Genet.* 1999; 3, 2-6.

[11] Hubbard, T. J. P. RMS/coverage graphs. A qualitative method for comparing three dimensional protein structure predictions. *Proteins Struct. Funct. Genet.* 1999; 3, 15-21.

[12] Rossman, M. G., & Argos, P., Exploring structural homology of proteins. *J. Mol. Biol.* 1976; 105, 75-95.

[13] Chothia, C., Lesk, A. M., The relation between the divergence of sequence and structure in proteins. EMBO J. 1986; 5, 823-826.

[14] Benner, S. A., Cannarozzi, G., Chelvanayagam, G. & Turcotte, M., *Bona fide* predictions of protein secondary structure using transparent analyses of multiple sequence alignments. *Chem. Rev.* 1997; 97, 2725-2843.

[15] Needleman, S. B. & Wunsch, C. D., A general method applicable to the search for similarities in the amino acid sequences of two proteins. *J. Mol. Biol.* 1970; 48, 443-453.

[16] Smith, T. F. & Waterman, M. S., Identification of common molecular subsequences. *J. Mol. Biol.* 1981; 147, 195-197.

[17] Thorne, J. L., Kishino, H. & Felsenstein, J., Inching toward reality. An improved likelihood model of sequence evolution. *J. Mol. Evol.* 1992; 34, 3-16.

[18] Cohen, M. A., Benner, S. A., Gonnet, G. H., Analysis of mutation during divergent evolution. The 400 by 400 dipeptide mutation matrix. *Biochem. Biophys. Res. Comm.* 1994; 199, 489-496.

[19] Benner, S. A., Cohen, M. A., Gonnet, G. H., Empirical and structural models for insertions and deletions in the divergent evolution of proteins. *J. Mol. Biol.* 1993; 229, 1065-1082.

[20] Gonnet, G. H., Benner, S. A. Computational Biochemistry Research at ETH. *Technical Report 154, Departement Informatik*, 1991 Swiss Federal Institute of Technology, Zurich, Switzerland.

[21] Gonnet, G. H., Cohen, M. A., Benner, S. A.,Exhaustive matching of the entire protein sequence database. *Science* 1992; 256, 1443-1445.

[22] Rost, B.; Sander, C., Prediction of protein secondary structure at better than 70-percent accuracy *J. Mol. Biol.* 1993; 32, 584-599.

[23] Sternberg, M. J. E., Taylor, W.R., Modeling the ATP binding site of onco-gene products, the epidermal growth-factor receptor and related proteins. *FEBS Lett.* 1984; 175, 387-392.

[24] Benner, S. A., Gerloff, D. L., Patterns of divergence in homologous pro-teins as indicators of secondary and tertiary structure. The catalytic domain of protein kinases. *Adv. Enz. Regul.* 1991; 31, 121-181.

[25] DeFay, T., Cohen, F. E., Evaluation of current techniques for ab initio protein structure preduction. *Proteins Struct. Funct. Genet.* 1995; 23, 431-445.

[26] Benner, S. A., Trabesinger-Rüf, N., Schreiber, D. R. Post-genomic science. Converting primary structure into physiological function. *Adv. Enzyme Reg.* 1998; 38, 155-180.

[27] Chircurel, M. Whatever happened to leptin? *Nature* 2000; 404, 538-540.

[28] Chandrasekharan, U. M., Sanker, S., Glynias, M. J., Karnik, S. S., Husain, A.,Angiotensin II forming activity in a reconstructed ancestral chymase. *Science* 1996; 271, 502-505.

[29] Liberles, D. S., Schreiber, D. R., Govindarajan, S., Chamberlin, S. G., Benner, S. A., The adaptive evolution database (TAED). *Genome Biol.* 2001; 2, 0003.1-0003.18

[30] Benner, S. A., Gerloff, D. L, Chelvanayagam, G., The phospho-β-galactosidase and synaptotagmin predictions. *Proteins. Struct. Funct. Genet.* 1995; 23, 446-453.

[31] Gerloff, D. L., Cohen, F. E., Korostensky, C., Turcotte, M., Gonnet, G. H., Benner, S. A., A predicted consensus structure for the N-terminal fragment of the heat shock protein HSP90 family. *Proteins Struct. Funct. Genet.* 1997; 27, 450-458.

[32] Tauer, A., Benner, S. A., The B12-dependent ribonucleotide reductase from the archaebacterium *Thermoplasma acidophila*. An evolutionary conundrum. *Proc. Natl. Acad. Sci. USA* 1997; 94, 53-58.

[33] Jenny, T. F., Gerloff, D. L., Cohen, M. A., Benner, S. A., Predicted secondary and supersecondary structure for the serine/threonine specific protein phosphatase family. *Proteins Struct. Funct. Genet.* 1995; 21, 1-10.

[34] Benner, S. A., Trabesinger-Ruef, N., Schreiber, D. R., Post-genomic science. Converting primary structure into physiological function. *Adv. Enzyme Regul.* 1998; 38, 155-180.

Chapter 3

OPTIMIZED NEEDLE BIOPSY STRATEGIES FOR PROSTATE CANCER DETECTION

Ariela Sofer

Department of Systems Engineering and Operations Research
George Mason University
asofer@gmu.edu

Jianchao Zeng

Imaging Science and Information Systems Center (ISIS)
Department of Radiology, Georgetown University Medical Center
zeng@isis.imac.georgetown.edu

Abstract   Clinical diagnosis of prostate cancer is most often done by transrectal ultrasound-guided needle biopsy. Because of the low resolution of ultrasound, however, the urologist cannot usually distinguish between cancerous and healthy tissue. Therefore, most biopsies follow standard procedures (known as "protocols") based on long-term physician experience. Recent studies indicate that these protocols may have a significant rate of false negative diagnoses. This research develops optimized biopsy protocols. We use real prostate specimens removed by prostatectomy to develop a 3D distribution map of cancer in the prostate. We develop also a probability model of the needle insertion procedure. Using this model, the tumor map, and the geometry of the biopsy needle, we obtain estimates for the probability of obtaining a positive biopsy in various zones of prostates with cancer. Using this we develop a nonlinear optimization problem that determines the protocols that maximize the probability of cancer detection for a given number of needles, and present new optimized protocols.

Keywords: Prostate cancer, optimization

Introduction

Prostate cancer is the most prevalent male malignancy and the second leading cause of death by cancer in American men. The American Cancer Society

P.M. Pardalos and J. Principe (eds.), Biocomputing, 45-58.
© 2002 *Kluwer Academic Publishers. Printed in the Netherlands.*

estimates that there will be about 198,100 new cases of prostate cancer in the United States in 2001, and about 32,000 men will die of the disease. Current screening for the cancer includes the prostate specific antigen (PSA) test and the digital rectal exam. However the cancer can only be correctly diagnosed by needle biopsy of the prostate and histopathology of the sampled tissues. The most common technique for detection of prostate cancer is transrectal ultrasound-guided (TRUS) needle core biopsy.

Since normal prostate tissue cannot usually be differentiated from cancerous tissue during the biopsy, a number of standard protocols have been developed to assist the urologist in performing the biopsy. A biopsy protocol designates the number of needles to be used used, and their location within the prostate. The most commonly used is the systematic sextant biopsy [[4]]. Recent studies [[1], [10]] have shown, however, that this strategy has an unacceptable level of false negative diagnoses, and that many patients who have a negative initial biopsy are found to have cancer in repeat biopsies.

As a result, recent clinical studies have investigated new protocols that have higher detection rates [[2], [3]]. The improvement in detection is obtained by adding up to seven more needles to the six of the sextant method in the biopsy.

Our approach is different: our goal is to develop optimized biopsy protocols. For a specified number of needles, an optimal protocol is one that maximizes the probability of detection of cancer in a patient. The hope is that with optimized protocols one could achieve improved detection rates with fewer needles.

Ideally, we would like to specify the precise locations within the pros tate where the needles should be placed. But because of the low resolution of ultrasound, it is not possible for the urologist to pinpoint the positions of the needles to a high degree of accuracy. This, for the purpose of biopsy guidance for the physician we have divided the prostate into a coarse grid of 48 anatomically-recognized zones within the prostate. Our proposed protocols will suggest to the physician in which zones the needles should be placed. The grid size of 48 is likely the upper limit, for which the zones are still distinguishable by the physician.

As the first step in determining the protocols that maximize the probability of detecting cancer, we develop a statistical distribution map of cancer in the prostate. The map is constructed from cancerous prostates that were removed via prostatectomy. Each of the prostates is first reconstructed into a 3D computerized model that accurately represent the anatomy of the prostate, and the distribution of cancer within it. Next, a fine grid (here, of 6000 grid points) is superimposed over the prostate model, and the presence of cancer in each grid point is identified. From this the 3D distribution map of tumor location is developed. Thus far, 301 prostates have been reconstructed and analyzed.

Using the distribution we next develop an optimization model that determines the location of a prescribed number of needles that maximizes the probability of detection of cancer. The model yields new biopsy protocols superior to the sextant method.

The paper is organized as follows: Section 1 discusses the reconstruction of the prostate models, and Section 2 discusses the construction of the statistical distribution map. The mathematical model for maximizing the probability of detecting cancer is presented in Section 3, and the optimization solution approach is discussed in Section 4. Section 5 presents the resulting biopsy protocols and compares them with the current procedures. Section 6 presents conclusions and further research.

1. Reconstruction of the prostate models

301 individual 3D prostate models were reconstructed from radical prostatectomy specimens. The original diagnosis of cancer for these patients occurred through a variety of detection methods: some cancers were diagnosed from biopsy using the sextant protocols, others only via repeat biopsies, some by finger-guided biopsy, and some by biopsy of suspicious areas appearing in the ultrasound images.

The reconstruction of each prostate model consists of the following steps [[13]]:

(a) Physical slicing and digitization. Each prostate specimen was sectioned in 4μm sections at 2.25mm intervals, and each slice was digitized with a scanning resolution of 1500 dots per inch.

(b) Extraction of key structures. Each digitized image was segmented by a pathologist to identify the key pathological structures, including surgical margins, capsule, urethra, seminal vesicle, and the tumor.

(c) Construction of a 3D frame model. The contours of each structure were identified on each slice, and then stacked up. Interpolation between adjacent pairs of contours was performed using a 3D elastic contour model [[8], [12]]. With this model the interpolation between contours C_1 and C_2 is performed by generating a force field that acts on C_1 and gradually forces it to move and conform to C_2. In the process, a specified number K of intermediate contours IC_j, $(j = 0, 1, \ldots K)$ are created between $C_1 = IC_0$ and $C_2 = IC_K$, resulting in a smoother transition between the two contours. The intermediate contours are defined by

$$IC_{j+1} = IC_j + F_j * IC_j, \quad j = 0, \ldots, K - 1.$$

Here F_j denotes the force field at intermediate stage j. It is mainly determined by the the Euclidian distance and the difference in orientation between pairs of line segments formed by selected points on C_1 and C_2.

The larger the distance and difference in orientation, the stronger the force will be.

(d) 3D surface reconstruction from the frame models. The 3D model of each structure in the prostate was finalized by tiling triangular patches onto the interpolated contours, using a deformable surface-spine model that employs a second-order partial differential equation to control the deformation of the surface. The reconstruction technique is described in detail in [Zeng et al.,1998a].

The successive stages in the reconstruction are shown in Figure 3.1.

2.   The Statistical distribution map

We now determine the statistical distribution map. The coarse grid 48-zone coarse grid used for biopsy guidance is illustrated in Figure 3.2. The grid has three transverse layers, base, mid, and apex. Each such layer is further divided into four coronal layers, which, following clinical conventions, are labeled from the posterior to the anterior as posterior 1, posterior 2, anterior 1, and anterior 2 respectively. Finally, the layers are divided from left to right into four sagittal layers, denoted as left lateral, left mid, right mid, and right lateral, respectively. It is noted that the size of the zones will vary with the size of a prostate model. A larger prostate model ends up with having a larger size for each of its 48 zones. This variation of zone sizes is the natural and consistent reflection of the original prostate, and it should not affect the accuracy of cancer distributions.

For determination of the statistical distribution map, each of the course-grid compartments is further divided to smaller subzones. We have used a total of 5^3 subzones per zone, so that the final grid superimposed over the prostate has 6000 subzones. The occurrence of cancer is calculated in each of the subzones for each of the patients; a subzone is considered positive for a given patient if it contains any part of a cancer. Subzones that contain no prostate tissue are marked accordingly.

Next, we estimate for each patient and for each of the coarse-grid zones, the probability that a needle biopsy in the zone will detect cancer. To ensure accuracy we will model the variability in the physician's placement of the needle. Specifically for each of the 48 zones we model (i) the depth of the needle insertion point (from apex to base); (ii) the longitudinal position of the needle insertion point (from posterior to anterior) and (iii) the firing angle of the needle. (Note that the physician controls the firing angle by rotating the ultrasound probe around its axis; there is only one degree of freedom, since the needle has a fixed angle with respect to the axis of the ultrasound probe.) We now assume that for each zone the three corresponding random variables controlling the needle trajectory are statistically independent Gaussian variables. (The parameters of these distributions will vary from zone to zone.) We then use a discretization

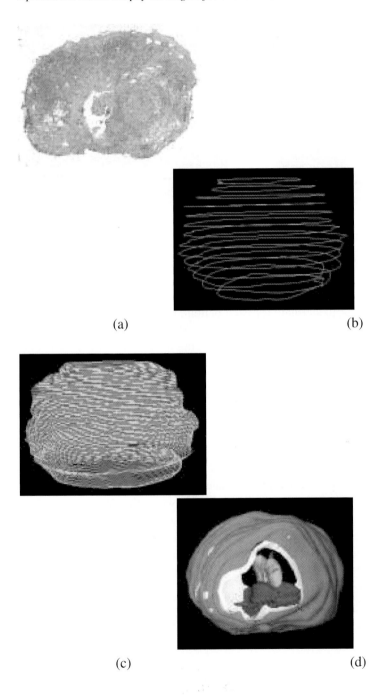

(a)

(b)

(c)

(d)

Figure 3.1. 3D reconstruction of prostate models: (a) Digitized image of a single slice of a sectioned prostate(b) Stacked surgical margin contour controls of original slices. (c) Surgical margin contour interpolation (d) Final 3-D reconstructed prostate model

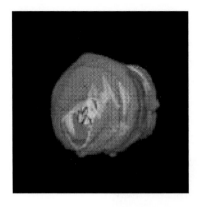
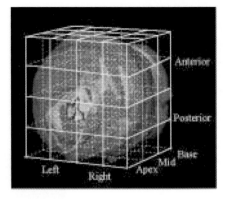

Figure 3.2.   (a) Reconstructed prostate model (b) 48-zone grid superimposed over prostate model.

of these distributions, to estimate for each patient, the probability that a needle probe in the zone will be positive. The analysis is based on the fine-grid cancer distribution map, the volume of each subzone, the geometry of the needle and the volume of the needle core.

3.   The Optimization problem

3.1.   Problem formulation

We now develop a mathematical model that determines the optimal locations for the biopsy needles based on the 3D cancer distribution model. The objective is to determine, for a specified number of needles, the protocol that maximizes the probability of detecting cancer. We note that while the detection rate increases with the number of needles, using optimized strategies could permit using fewer needles, thus avoiding patient discomfort and saving costs.

To formulate the problem, let n denote the number of zones in the coarse grid (here $n = 48$), and let m denote the number of patients in our sample ($m = 301$). Define variables $x_j, j = 1, \ldots, n$, by

$$x_j = \begin{cases} 1 & \text{if a biopsy is taken in zone } j, \\ 0 & \text{otherwise.} \end{cases}$$

Let p_{ij} be the estimated probability that a needle in zone j will detect cancer in patient i ($i = 1, \ldots, m, j = 1, \ldots, n$), and let $q_{ij} = 1 - p_{ij}$. We have that

$$(q_{ij})^{x_j} = \begin{cases} 1 & \text{if } x_j = 0 \\ q_{ij} & \text{if } x_j = 1. \end{cases}$$

Now in general, the occurrence of cancer in adjacent prostate zones is correlated. But for a given patient i in our sample the location of cancer is known, hence the occurrence of cancer is no longer a random variable (but its realization). In contrast, the outcome of a needle biopsy in a given zone for this patient is a random variable. The randomness is due only to the uncertainty in how the physician places the needle; if the placement of the needle were totally deterministic, the outcome of the needle biopsy would also be deterministic. Since the physician's positioning of different needles can be assumed to be statistically independent, the outcomes of needle biopsies in different zones can be assumed to be statistically independent. Thus the probability that a set of needle biopsies diagnoses cancer in patient i of our sample is

$$1 - \prod_{j=1}^{n}(q_{ij})^{x_j}.$$

It follows that the k-needle biopsy protocol that maximizes the probability of detection for a randomly selected patient in our sample solves

$$\text{maximize} \quad \frac{1}{m}\sum_{i=1}^{m}(1 - \prod_{j=1}^{n}(q_{ij})^{x_j})$$

$$\text{subject to} \quad \sum_{j=1}^{n}x_j = k \tag{1}$$

$$x_j \in \{0, 1\}.$$

Note that when scaled by m, the objective function is the expected number of patients in the sample to be diagnosed with cancer. Problem (1) can be written equivalently as

$$\text{minimize} \quad \sum_{i=1}^{m}\prod_{j=1}^{n}(q_{ij})^{x_j}$$

$$\text{subject to} \quad \sum_{j=1}^{n}x_j = k \tag{2}$$

$$x_j \in \{0, 1\}.$$

Additional constraints reflecting physician preferences can also be added. For example, left-right symmetry of the protocols is obtained by the constraints $x_l = x_r$ for every pair of zones l and r that are left-right symmetric. One may also restrict biopsies to zones in the posterior area (since the anterior is more difficult to probe), by imposing the constraint $x_a = 0$ for all zones a in the anterior. For simplicity we will let X denote the feasible set—the set of vectors $x \in \{0, 1\}$ that satisfy $\sum_{j=1}^{n}x_j = k$ and any additional desired constraint.

The optimization problem (2) is a nonlinear integer problem. In its current form the problem is difficult to solve, since its objective is a nonconvex function in the integer variables x_j. However we can transform it to a more tractable problem. Let ϵ be some (sufficiently) small positive tolerance (say, $\epsilon = 10^{-10}$), and define the $m \times n$ matrix U by $u_{ij} = -\log(\max\{q_{ij}, \epsilon\})$. Denoting $z_i = \sum_{j=1}^{n} u_{ij}x_j$, the optimization problem is equivalent to the problem

$$
\begin{aligned}
\text{minimize} \quad & \sum_{i=1}^{m} e^{-z_i} \\
\text{subject to} \quad & z - Ux = 0 \\
& x \in X.
\end{aligned}
\tag{3}
$$

Although the resulting problem is still nonlinear and integer, the transformed form is more convenient, since the objective function is convex.

3.2.   Solution of the optimization problem

To solve problem (3) we use a generalized decomposition algorithm [[6], [5]] framework. The idea is to decompose the problem in a way that enables the creation of a sequence of easier subproblems that gradually provide tighter lower and upper bounds on the optimal objective. Here this is done as follows: We rewrite (3) as

$$
\underbrace{\begin{aligned}
\underset{x \in X}{\text{minimize}} \quad & \underset{z}{\min} & g(z) = \sum e^{-z_i} \\
& \text{subject to} & z - Ux = 0
\end{aligned}}_{\text{primal problem}}
\tag{4}
$$

Let $L(x, \lambda) = \sum_{i=1}^{m} e^{-z_i} - \lambda^T(z - Ux)$ be the Lagrangian for the primal problem. Now because the problem is convex, its objective is equal to that of its Lagrangian dual [[9]]. Hence (4) is equivalent to

$$
\underset{x \in X}{\text{minimize}} \quad \underset{\lambda}{\max} \quad \underset{z}{\min} \quad \left(\sum e^{-z_i} - \lambda^T(z - Ux) \right) .
\tag{5}
$$

Problem (5) can be written in equivalent form

$$
\begin{aligned}
\underset{x \in X, \delta}{\text{minimize}} \quad & \delta \\
& \delta \geq \underset{z}{\min} \left(\sum e^{-z_i} - \lambda^T(z - Ux) \right) \quad \forall \lambda
\end{aligned}
\tag{6}
$$

Suppose we start from an initial feasible integer point x^0. It is easy to show that for a given vector x^t, the solution to the primal problem in (4) is

$$
z^t = Ux^t \quad \lambda_i^t = -e^{-z_i^t}.
$$

This yields at iteration t an upper bound UB=$\min\{UB, g(x^t)\}$ on the optimal objective, where at iteration 0, UB=$g(z^0)$. Now given the iterates $x^0, \ldots, x^t$,

Table 3.1. Estimated detection rates with optimized symmetric protocols using 6, 8, and 10 needles. Estimated detection rate for sextant method is 67.3%

No. of Needles	Posterior Only	Entire Gland
6	78.8%	79.3%
8	81.6%	82.9%
10	84.2%	85.5%

the problem

$$\underset{x \in X, \delta}{\text{minimize}} \quad \delta$$

$$\delta \geq (\lambda^j)^T U x + \sum e^{-z_i^j} - (\lambda^j)^T z^j, \quad j = 0, \ldots, t$$

is a "relaxation" of Problem (6), in the sense that it relaxes its constraints. Thus the solution provides a lower bound on the optimal objective LB=δ and a new starting x^{t+1} for the primal.

We can thus obtain a sequence of solutions to the primal problem and the dual problem, with the former yielding a nonincreasing sequence of upper bounds to the optimal objective value, and the latter yielding a nondecreasing sequence of lower bounds to this objective value. The algorithm terminates when the upper and lower bounds differ by less than a prescribed tolerance. The relaxed dual problems are linear programs with integer variables, and are easily solved by the software package ILOG CPlex 6.5 [[7]].

4. Optimized protocols

We have obtained preliminary results using the sample of 301 prostate analyzed so far. The resulting estimated detection rates for optimized 6-, 8-, and 10-needle biopsy protocols are shown in Table 3.1. As one can see, the estimated detection rate for our optimal 6-needle protocols is about 79%. In contrast, the estimated detection rate for the sextant method is about 67%. Thus it is possible to improve detection rates with 6 needles only, just by using optimized protocols. Additional needles will further improve the detection rates as indicated in the Table. The corresponding optimal biopsy protocols are shown in Table 3.2.

5. Conclusions and future work

Our Preliminary results show that the optimal biopsy protocols have a substantial improvement over the protocols currently used clinically. The next step in our research will be to evaluate the new biopsy protocols against the conventional approaches via simulated virtiual biopsy.

Our research group [[13]] has used the reconstructed model t0 develop a 3-D visualization and simulation system. The system is developed using C++ and

Table 3.2. Optimal symmetric biopsy protocols for 6, 8, and 10 needles. x indicates a zone that is part of the biopsy; ll, lm, rm, and rl indicate left lateral, left mid, right mid, and right lateral respectively; p1, p2, a1, and a2 denote posterior 1 and 2, and anterior 1 and 2 respectively.

		Base				Mid				Apex			
		ll	lm	rm	rl	ll	lm	rm	rl	ll	lm	rm	rl
6 needles	a2												
	a1					x			x				
	p2										x	x	
	p1					x			x				
8 needles	a2												
	a1					x			x				
	p2	x			x						x	x	
	p1					x			x				
10 needles	a2												
	a1					x			x				
	p2	x			x						x	x	
	p1					x			x		x	x	

the object-oriented 3-D visualization development toolkit Open-Inventor on an SGI Onyx Workstation. Graphical user interface is realized based on the Motif toolkit. While menu operations are mostly performed using a two-dimensional (2-D) mouse, the interactive biopsy simulation is mainly carried out using a 6 degrees of freedom tracking device which is especially integrated in the visualization system. In addition to general visualization functions, such as model manipulation (e.g., rotation, translation and zooming) and model property change (e.g., transparency and color), this system primarily provides functions specific for the prostate needle biopsy. It has two simulation modes: an automatic simulation and an interactive simulation. The whole process of a prostate needle biopsy with any specific scheme can be simulated based on the reconstructed 3-D prostate surface models. In the automatic simulation mode, the locations for needle insertion on the surface of the prostate are calculated automatically by the computer based on the requirement of the specific biopsy scheme. Needles are then mounted to the positions in the calculated poses. After shooting the needles, the system then detects which needles will hitting the tumor(s) inside the prostate by calculating ray intersection along the needle direction with the tumors. If the biopsy is positive, the system calculates the positive needle core volumes by the amount of intersection and displays the results on the screen. Each step of the automatic biopsy simulation process can also be visualized from any perspective by manipulating the 3-D prostate model in real time.

For the interactive simulation, a 6 degree of freedom tracking device is integrated to simulate the ultrasound probe used during actual prostate biopsy procedure. The tracking device consists of an ultrasound transmitter, a controller, and a freely movable receiver device that serves as a tracker. With this device, the system can track both the position and the orientation angles (pitch, yaw, roll) of the receiver in real time. The tracking information is simultaneously used in controlling movement of the virtual ultrasound probe in the visualization system. The synthesized ultrasound images are refreshed in real time to follow the movement of the probe. The ultrasound images show intersectional anatomical slices of the prostate as biopsy guidance for the user (a urologist). With this interactive simulation mode, the urologist can perform a virtual needle biopsy as though he/she is performing a real biopsy on a patient. He/she determines the location for each needle insertion based on the specified biopsy scheme under the guidance of the synthesized ultrasound image. The angle of the needle is fixed with the ultrasound probe, and the upcoming path of the needle is always displayed and overlaid on the ultrasound image so that the urologist knows where the needle will go through inside the prostate. The result of a biopsy is automatically calculated by the system after each biopsy and is displayed to tell the urologist whether the biopsy is positive or negative and how much the positive needle core volume is. Figure 3.3a shows the virtual ultrasound probe and the needle in use, while Figure 3.4b shows the needles after being fired into the prostate.

We are currently reconstructing 100 additional models. A virtual biopsy will be performed by a urologist on these 100 new models to evaluate the new protocols against current practices. Pending the outcome of these simulations, the subsequent step will be clinical trials.

Acknowledgments

Ariela Sofer is partially supported by National Science Foundation grant DMI-9800544. Jianchao Zeng is supported in part by The Whitaker Foundation Biomedical Engineering Program grant RG-99-0115. We wish to acknowledge the input of John J. Bauer (WRMC), Wei Zhang and Isabell A. Sesterhenn (AFIP), and Judd W. Moul (CPDR), Brett Opell, and Seong K. Mun,

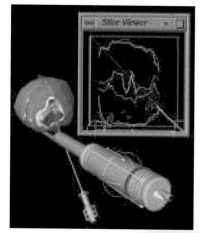

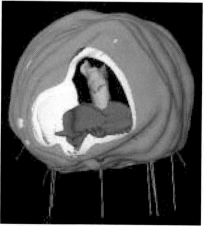

Figure 3.3a.   Virtual ultrasound probe and
needle in use.

Figure 3.4b.   Needles after being fired into
prostate

References

[1] Bankhead C. (1997). "Sextant biopsy helps in prognosis of Pca, but its not foolproof," *Urology Times* Vol. 25, No. 8. (August).

[2] Chang J.J., Shinohara K., Bhargava V., Presti, J.C. Jr. (1997). "Prospective evaluation of lateral biopsies of the peripheral zone for prostate cancer detection," *J. Urology* Vol. 160, pp. 2111–2114.

[3] Eskew, A.L., Bare,R.L., McCullough D.L. (1997). "Systematic 5-region prostate biopsy is superior to sextant method for detecting carcinoma of the prostate." *J. Urology* Vol. 157, pp. 199–202.

[4] Hodge K.K, McNeal J.E., Terris M.K., and Stamey T.A. (1989). "Random systematic versus directed ultrasound guided trans-rectal core biopsies of the prostate," *J. Urology* Vol. 142, pp. 71–74

[5] Floudas, C.A. (1995). *Nonlinear and Mixed-Integer Ptimization*, Oxford Univesity Press.

[6] Geoffrion A.M (1972). "Generalized Bender's decomposition," *J. Optim. Theory and its Appl.*, 10, pp. 237–253.

[7] ILOG CPLex 6.5 User Manual, ILOG, 1999.

[8] Lin W., Liang C., Cheng C. (1988). "Dynamic elastic interpolation for 3D medical inage reconstruction from serial cross section." *IEEE Transactions Medical Imaging* 7, pp. 225–232.

[9] Nash, S.G, and Sofer, A (1996). *Linear and Nonlinear Programming*, McGraw Hill.

[10] Rabbani F., Stroumbakis N., Kava B.R., Cookson M.S., and Fair W.R. (1998). "Incidence and clinical significance of false-negtive sextant prostate biopsies," *J. Urology* Vol. 159, pp. 1247–1250.

[11] Sofer, A., Zeng, J., Opell, B., Bauer, J., Mun, S.K. (2000). "Optimal Biopsy Protocols for Prostate Cancer," submitted for publication.

[12] Xuan, J. Sesterhenn I., Hayes WS, Wang Y., Adali T., Yagi Y, Freed-
man M.T, Mun S.K. (1998). "Surface Reconstruction and Visualization
of the surgical prostate model" *Preceedings of the SPIE Medical Imaging
Conference*, 1997, 3031: 50–61.

[13] Zeng, J., Bauer J.J., Yao X., Zhang W., Sesterhenn I.A., Connelly R.R.,
Moul J., and Mun S.K. (2000). "Building an accurate 3D map of prostate
cancer using computerized models of 280 whole-mounted radical prostatectomy specimens." *Proc. of SPIE Medical Imaging Conference,* Vol.
3976, pp. 466-477.

[14] Zeng, J., Bauer, J., Sofer, A., Yao, X., Opell, B., Zhang, W., Sestrehenn,
I. A., Moul, J. W., Lynch, J., Mun, S. K. (2000). Distribution of Prostate
Cancer for Optimized Biopsy Protocols. In it Proceedings of the Medical
Image Computing and Computer Assisted Intervention Conference pp.
287-296.

Chapter 4

PHASE ENTRAINMENT AND PREDICTABILITY OF EPILEPTIC SEIZURES *

L.D. Iasemidis
Bioengineering
Center for Systems Science and Engineering Research,
Arizona State University
leon.iasemidis@asu.edu

D.-S. Shiau
Statistics
University of Florida
shiau@epilepsy.health.ufl.edu

P. Pardalos
Center for Applied Optimization
Industrial and Systems Engineering
University of Florida
pardalos@ufl.edu

J.C. Sackellares
Neurology; Bioengineering; Neuroscience
University of Florida
sackellares@epilepsy.health.ufl.edu

Abstract   Epilepsy is one of the most common disorders of the nervous system, second only to strokes. We have shown in the past that progressive entrainment between an epileptogenic focus and normal brain areas results to transitions of the

*This research is supported by NIH, NSF, VA, Whitaker and DARPA research grants.

P.M. Pardalos and J. Principe (eds.), Biocomputing, 59-84.
© 2002 *Kluwer Academic Publishers. Printed in the Netherlands.*

brain from chaotic to less chaotic spatiotemporal states, the well-known epileptic seizures. The entrainment between two brain sites can be quantified by the T-index between measures of chaos (e.g., Lyapunov exponents) estimated from the brain electrical activity (EEG) at these sites. Recently, by applying optimization theory, and in particular quadratic zero-one programming, selecting the most entrained brain sites 10 minutes before seizures and subsequently tracing their entrainment backward in time over at most 2 hours, we have shown that over 90% of the seizures in five patients with multiple seizures were predictable [23]. In this communication we show that the above procedure, applied to measures of angular frequency in the state space (average rate of phase change of state) estimated from EEG data per recording brain site over time in one of our patients with 24 recorded seizures, produces very similar results about the predictability of the epileptic seizures (87.5%). This finding implies an interrelation of the phase and chaos entrainment in the epileptic brain and may be used to refine procedures for long-term prediction of epileptic seizures as well as to generate a model of the disorder within the framework of dynamical nonlinear systems.

1.   Introdution

Epilepsy is characterized by recurrent paroxysmal electrical discharges of the cerebral cortex that result in intermittent disturbances of brain function. It affects approximately 1% of the population. For some types of epilepsy (e.g., focal or partial epilepsy), there are structural changes in neuronal circuitry within localized regions of the cerebral cortex. These abnormal regions produce intermittent organized quasi-rhythmic discharges. These discharges then spread from the region of origin (epileptogenic focus) to activate other areas of the cerebral hemispheres. The hippocampus is the most common location of an epileptogenic focus. The structural abnormalities in the epileptogenic hippocampus include neuronal loss, dendritic simplification, and axonal sprouting of dentate granule cells [11, 32, 34, 4]. Also, there are metabolic abnormalities in epileptogenic regions, even during the interictal (between seizures) state. Localized zones of hypometabolism are detectable with PET scans [10, 2, 38]. While the macroscopic and microscopic features of the epileptogenic zone have been described, the mechanism by which these fixed disturbances in local circuitry produce intermittent disturbances of brain function is not understood yet.

Traditionally, the initial occurrence of the characteristic focal rhythmic EEG discharge is considered to be the onset of a seizure. However, through analysis of the spatiotemporal dynamics of such EEG recordings in patients with medically intractable temporal lobe epilepsy, we were the first to discover a preictal transition that precedes seizures for periods on the order of minutes to hours [17, 18, 20, 19, 21]. This preictal dynamical transition is characterized by a progressive convergence (entrainment) of measures of chaos (maximum Lyapunov exponents, i.e. L_{max} values) at specific anatomical areas. Although the existence of the preictal transition period has recently been confirmed by other groups [8, 28, 31, 30], the characterization of this spatiotemporal transition is still far from complete and therefore, the development of a model for the mechanism of generation of epileptic seizures remains a difficult task. For example, we have shown that: 1) even in the same patient, different set of brain sites may be entrained from one seizure to the next; 2) In addition to the widespread nature of this entrainment, resetting of the entrainment of the normal sites with the epileptogenic focus (critical brain sites) follows the end of each seizure [24]. Therefore, it is expected that complete or partial resetting of the observed preictal entrainment in the epileptic brain after the occurrence of a clinical or a subclinical seizure affects the route of the brain toward a subsequent seizure.

In this chapter, a new measure, the rate of change of the phase of a state (angular frequency Ω_{max}) in the state space of each brain site, is proposed for additional quantification of the preictal transition. The definition and the method of estimation of this novel measure from the EEG is described in section 2. The selection of specific brain areas using optimization theory in order to maximize the detection of the preictal transition is addressed in section 3. In section 4, the method to determine the predictability of an epileptic seizure is presented. Predictability results from the application of the new measure Ω_{max} as well as of the old measure L_{max} of chaos to the EEG are comparatively given in section 5. These results are discussed in the final section 6.

2. Nonlinear dynamical measures

2.1. Measures of chaos (STL_{max})

Since its discovery by Richard Caton [5] and its first systematic investigation by Hans Berger [3, 12], the electroencephalogram (EEG) has been the most utilized signal to clinically assess the brain function. Unfortunately, traditional signal processing theory, based on very simple assumptions about the system that produces the signal (e.g. linearity assumption) has met the challenge of quantification of EEG with varying degrees of success. This limitation stems from the fact that the EEG is generated by a nonlinear system, the brain. EEG characteristics such as alpha activity and seizures, instances of bursting behavior during light sleep, amplitude dependent frequency behavior (the smaller the amplitude the higher the EEG frequency) and existence of frequency harmonics (e.g. under photic driving conditions) are typical features of the EEG signal. These characteristics all belong to the long catalog of properties of typical nonlinear systems [25]. The EEG, the output of a non-stationary multidimensional system, has statistical properties that depend on both time and space [29]. Nonlinear components of the brain (neurons) are densely interconnected and the EEG recorded from one site is inherently related to the activity at other sites.

These components may functionally interact at different time instants. This makes the EEG a multivariable, nonlinear, non-stationary time series.

A well-established technique for visualizing the dynamical behavior of a multivariable system is to generate a state space portrait of the system. A state space portrait is created by treating each time-dependent variable of the system as a component of a vector in a multidimensional space, called state space of the system. Each vector in the state space represents an instantaneous state of the system. These time-dependent vectors are plotted sequentially in the state space to represent the evolution of the state of the system over time. For many systems, this graphic display illustrates an object confined over time to a sub-region of the phase space. Such sub-regions of the state space are called "attractors". The geometrical properties of these attractors provide information about the steady states of the system.

One of the problems in analyzing multidimensional systems is to decide which observables of the system (variables that can be measured) to analyze. Experimental constraints may limit the number of observables. However, when the variables of the system are related over time, which must be the case for any dynamical system to exist as a system, proper analysis of a single observable (e.g. EEG recorded from one electrode) can provide information about the variables of the system (e.g. activities at other electrode sites) that are related to this observation. Thus, it is possible to understand important features of a dynamical system through analysis of a single observable over time.

In principle, through the method of delays described by Packard et al. [36] and Takens [40], sampling of a single observable over time can approximate the position (state) of the system in a space spanned by the system variables related to this observable. Sampling with the method of delays can be used to reconstruct a multidimensional state space from a single-channel EEG signal. In such an embedding, each state is represented in the state space by a vector $X(t)$ whose components are the delayed versions of the original single-channel EEG time series $u(t)$, that is:

$$x(t) = [u(t), u(t - \tau), ..., u(t - (p - 1) \times \tau)],$$

where $x(t)$ is a vector in the state space at time t, τ is the time delay between successive components of $x(t)$, and p is the embedding dimension of the reconstructed state space. The embedding dimension p is the dimension of the state space that contains the steady state of the system (i.e. attractor) and it is always a positive integer. On the other hand, the attractors' dimension D may be a positive non-integer (fractal). D is directly related to the number of variables of the system and is usually inversely related to the existing coupling among them. According to Takens, the embedding dimension p should be at least equal to $(2D + 1)$ in order to correctly embed an attractor in the state space. Of the many different methods used to estimate D of an object in the state space, each has its own practical problems [9, 13]. The measure most often used to estimate D is the state space correlation dimension ν. Methods for calculating ν from experimental data have been described [33, 27] and were employed in our work to approximate D of the epileptic attractor. In the EEG data we have analyzed to date, ν is found to be between 2 and 3 during an epileptic seizure. Therefore, in order to capture characteristics of the epileptic attractor, we have used an embedding dimension p of 7 for the reconstruction of the state space.

An attractor is chaotic if, on the average, orbits originating from similar initial conditions (nearby points in the phase space) diverge exponentially fast (expansion process). If these orbits belong to an attractor of finite size, they will fold back into it as time evolves (folding process). The result of these two processes may be a stable, topologically layered, attractor [14]. When the expansion process, on average, overcomes the folding process in some eigendirections of the attractor, the attractor is called chaotic. The measures that quantify the chaoticity of an attractor are the Kolmogorov entropy (K) [26] and the Lyapunov exponents, typically measured in bits/sec [35, 37, 42]. For an attractor to be chaotic, the Kolmogorov entropy or at least the maximum Lyapunov exponent (L_{max}) must be positive. The Kolmogorov entropy (K), measures the uncertainty about the future state of the system given information about its previous states in the state space. The Lyapunov exponents measure this average uncertainty along the local eigenvectors of an attractor in the state space. If the state space is of p dimensions, we can estimate theoretically up to p Lyapunov exponents. However, as expected, only ($[D] + 1$) of them will be real. The rest will be spurious [1]. Methods for calculating these dynamical measures from experimental data have been published [1, 43, 16, 7]. The estimation of the largest Lyapunov exponent in a chaotic system has been shown to be more reliable and reproducible than the estimation of the remaining exponents [13, 41], especially when D is unknown and changes over time, as it is the case with high-dimensional and nonstationary data.

Since the brain is a nonstationary system, algorithms used to estimate measures of the brain dynamics should be capable of automatically identifying and appropriately weighing existing transients in the data. The method we developed for estimation of L_{max} for nonstationary data, called STL (Short Time Lyapunov), considers possible nonstationarities in the EEG. This method is explained in detail in Iasemidis et al. [18, 16]. We apply the STL algorithm to EEG tracings from electrodes in multiple brain sites, to create a set of STL_{max} time series. This set of time series contains local (in time and in space) information about the brain as a dynamical system. It has been shown that it is at this level of spatiotemporal analysis that reliable detection of the transition to epileptic seizures, long before they actually occur, is derived.

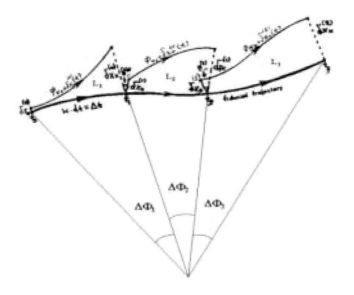

Figure 4.1. Diagram to illustrate the estimation of STL_{max} and Ω_{max} measures in the state space.

The largest Lyapunov exponent (L_{max} or L_1) is defined as the average of local Lyapunov exponents L_{ij} in the state space, that is:

$$L_{max} = \frac{1}{N_\alpha} \cdot \sum_\alpha L_{ij},$$

where N_α is the total number of the local Lyapunov exponents that are estimated from the evolution of adjacent points (vectors) in the state space, $X_i = X(t_i)$ and $X_j = X(t_j)$, according to:

$$L_{ij} = \frac{1}{\Delta t} \cdot \log_2 \frac{|X(t_i + \Delta t) - X(t_j + \Delta t)|}{|X(t_i) - X(t_j)|}$$

where Δt is the evolution time allowed for the vector difference $\delta_0(X_{ij}) = |X(t_i) - X(t_j)|$ to evolve to the new difference $\delta_k(X_{ij}) = |X(t_i + \Delta t) - X(t_j + \Delta t)|$, $\Delta t = k \cdot dt$ with dt the sampling period of the data $u(t)$ (see Figure 1). If Δt is given in sec, L_{max} is in bit/sec.

2.2. Measure of phase / angular frequency (Ω_{max})

The difference in phase ($\Delta\Phi$) in the state space is defined as the average of the local phase differences $\Delta\Phi_i$, that is:

$$\Delta\Phi = \frac{1}{N_\alpha} \cdot \sum_{i=1}^{N_\alpha} \Delta\Phi_i$$

where N_α is the total number of phase differences estimated from the evolution of $X(t_i)$ to $X(t_i + \Delta t)$ in the state space, and $\Delta\Phi_i$ is given by (see also Figure 1):

$$\Delta\Phi_i = |\arccos(\frac{X(t_i) \cdot X(t_i + \Delta t)}{\| X(t_i) \| \cdot \| X(t_i + \Delta t) \|})|$$

Then, the angular frequency Ω_{max} is:

$$\Omega_{max} = \frac{1}{\Delta t}\Delta\Phi$$

If Δt is given in sec, then Ω_{max} is in rad/sec.

2.3. Statistical distances among chaos and phase spatiotemporal profiles

We employ the T-index (from the well-known t-test for comparisons of means of paired –dependent- observations) as a measure of distance between the mean values of pairs of STL_{max} or Ω_{max} profiles over time. The T-index at time t between the Ω_{max} profiles of electrode sites i and j is then defined as:

$$T_{ij}(t) = \frac{|E\{\Omega_{max,i}(t) - \Omega_{max,j}(t)\}|}{\sigma_{ij}(t)/\sqrt{N}}$$

where $E\{\cdot\}$ denotes the average of all differences $\Omega_{max,i}(t) - \Omega_{max,j}(t)$ within a moving window $w_t(\lambda)$ defined as:

$$w_t(\lambda) = \begin{cases} 1 & \text{if } \lambda \in [t - N - 1, t] \\ 0 & \text{if } \lambda \notin [t - N - 1, t], \end{cases}$$

where N is the length of the moving window and $\sigma_{ij}(t)$ is the sample standard deviation of the Ω_{max} differences between electrode sites i and j within the moving window $w_t(\lambda)$. The thus defined T-index follows a t-distribution with $N - 1$ degrees of freedom.

In the estimation of the $T_{ij}(t)$ indices in our data we used N=60 (i.e., averages of 60 differences of STL_{max} exponents between a pair of electrode sites per moving window). Since each value in the STL_{max} or Ω_{max} profiles is derived from a 10.24 second EEG data segment, the length of the window used corresponds to approximately 10 minutes in real time units. Therefore, a two-tailed t-test with N-1=59 degrees of freedom, at a statictical significance level α should be used to test the null Hypothesis H_0 = 'brain sites i and j acquire identical STL_{max} or Ω_{max} values at time t'. If we allow $\alpha = 0.1$, the probability of a type I error (the probability of falsely rejecting H_0 if H_0 is true), is 10%, or better. For the T-index to accept H_0 (i.e., statistically claim that electrode sites i and j have the identical STL_{max} or Ω_{max} values at time t), with the 90% confidence level, $T_{ij}(t)$ should be less than 1.671. If we set $\alpha = 0.2$ for the same test, $T_{ij}(t)$ should be less than 1.296 to accept H_0 with 80% confidence level.

3. Selection of brain sites: optimization

For many years the Ising model [6, 39] has been a powerful tool in studying phase transitions in statistical physics. Such an Ising model can be described by a graph $G(V, E)$ having n vertices $\{v_1, \ldots, v_n\}$ and each edge $(i, j) \in E$ having a weight (interaction energy) J_{ij}. Each vertex v_i has a magnetic spin variable $\sigma_i \in \{-1, +1\}$ associated with it. An optimal spin configuration of minimum energy is obtained by minimizing the Hamiltonian

$$H(\sigma) = - \sum_{1 \leq i \leq j \leq n} J_{ij}\sigma_i\sigma_j \text{ over all } \sigma \in \{-1, +1\}^n.$$

This problem is equivalent to the combinatorial problem of quadratic bivalent programming [15].

Motivated by the application of the Ising model to phase transitions we have used quadratic bivalent (zero-one) programming for the optimal selection of brain sites at periods prior to brain phase transition (epileptic seizures) [22, 23]. The objective function to be minimized is the distance of measures of chaos (STL_{max}) and/or of angular frequency (Ω_{max}) between recording brain sites. These measures were estimated as described before. The sites selected by the optimization method have provided two important insights. First, sites participating in the preictal transition could thus be identified. It was observed that although these sites differ from seizure to seizure, the sites that were most frequently selected were located in the epileptogenic zone. Second, convergence of dynamical measures of the selected sites over time, a phenomenon we have called dynamical entrainment, could be detected well before the onset of an impending epileptic seizure.

More specifically, we considered the integer 0-1 problem:

$$\min (x'\boldsymbol{T}x) \text{ with } x \in \{0, 1\}^n \text{ subject to the constraint } \sum_{i=1}^{n} x_i = k, \qquad (1)$$

where n is the total number of electrode sites and k the number of sites to be selected. The elements of the matrix $\boldsymbol{T} = (\boldsymbol{T}_{ij})$ are statistical measures of the distances of brain sites i and j with respect to the estimated mean and standard deviation of their STL_{max} and/or Ω_{max} values within 10 minute windows W. The statistical measures of distance we have used in this analysis are the T-indices from the well known t-test in statistics and are described in the previous section. If we include the constraint (1) in the objective function $f(x) = x'\boldsymbol{T}x$ by introducing the penalty

$$\mu = \sum_{j=1}^{n} \sum_{i=1}^{n} \boldsymbol{T}_{ij} + 1,$$

the optimization problem becomes equivalent to an unconstrained global optimization problem:

$$\min \left[x'\boldsymbol{T}x + \mu \left(\sum_{i=1}^{n} x_i - k \right)^2 \right], \text{ where } x \in \{0, 1\}^n \qquad (2)$$

The electrode site i is selected if $x_i = 1$ in the solution $\boldsymbol{x} = (x_1, \ldots, x_n)^T$ of (2). Led by a variety of empirical correlations and numerical experiments we have chosen to set the value of $k = 5$ (i.e., to select the 5 most entrained electrode sites) as a balance between sensitivity and specificity. Higher values of k decreased specificity whereas lower values of k decreased the sensitivity of the algorithm.

4. Predictability analysis

Figure 2 shows the 28 electrode montage used in our laboratory for subdural and depth EEG recordings. In this communication we report results from the analysis of continuous EEG signals in one patient with 24 epileptic seizures for at most 2 hour before to 2 hour after a seizure. The EEG was sampled with a sampling frequency of $200Hz$ and was band-pass filtered at $[0.1, 70Hz]$. By dividing the recorded EEG data from an electrode site into sequential non-overlapping segments, each 10.24 sec in duration, and estimating STL_{max} and Ω_{max} for each of these segments, profiles of STL_{max} and Ω_{max} over time are generated. Through the application of the quadratic zero-one optimization formulation described in the previous section, a tuple of five electrode sites ($k = 5$), which produce the minimal average of T-indices over all possible electrode pairs in the tuple, is selected separately per STL_{max} and Ω_{max} profiles. After the selection, the average T-index curve among those five sites is reconstructed backward in time up to 2 hours before the seizure to determine whether that seizure was predictable.

For illustration purposes, the smoothed (10-minute moving average) STL_{max} and Ω_{max} profiles of the five optimally selected electrodes, for three sets of 3

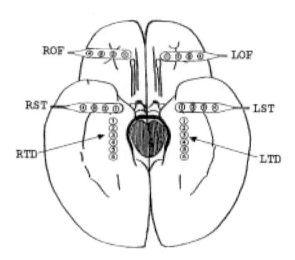

Figure 4.2. Diagram of the depth and subdural electrode placement. Electrode strips are placed over the left orbitofrontal (LOF), right orbitofrontal (ROF), left subtemporal (LST) and right subtemporal cortex. Depth electrodes are longitudinally placed in the left temporal (LTD) and right temporal (RTD) hippocampus to record bilateral hippocampal EEG activity.

successive seizures in each set, are shown in Figures 3, 4 and 5 respectively. The optimal electrodes were selected in a 10 minute interval prior to the second seizure of each set. For each set of seizures, STL_{max} and Ω_{max} profiles clearly converge (entrain) before the second seizure and either both (see Figure 3) or one of them (see Figures 4 and 5) diverge (disentrain) in this seizure's postictal period. The average T-index curves that quantify this preictal entrainment and postictal disentrainment among the selected electrodes for each of the corresponding 3 sets of seizures are respectively shown in Figures 6, 7 and 8. The second and third sets of seizures were included herein to show that STL_{max} and Ω_{max} measures are not identical in the detection of the entrainment and disentrainment transition across epileptic seizures in the same patient.

Based on the T-index curve described above, the decision of whether a seizure is predictable is determined by the following steps:

1. Starting with the first T-index value before a seizure's onset, we sequentially average the T-index values moving backwards in time and find the first time point where the averaged T-index value is greater than the critical value T_α extracted from the t-distribution for a given significance level α. Then, the duration of the entrainment (i.e., the preictal transition period PTP) can be defined as the time interval from the previously identified time point to the seizure onset time point.

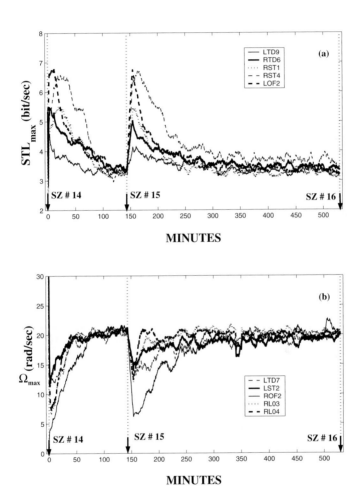

Figure 4.3. (a) Smoothed STL_{max} profiles of the 5 optimally selected electrodes over time (including seizures 14, 15, and 16). The optimal electrodes were selected 10 minutes before seizure 15. (b) Smoothed Ω_{max} profiles from the same EEG data as in (a).

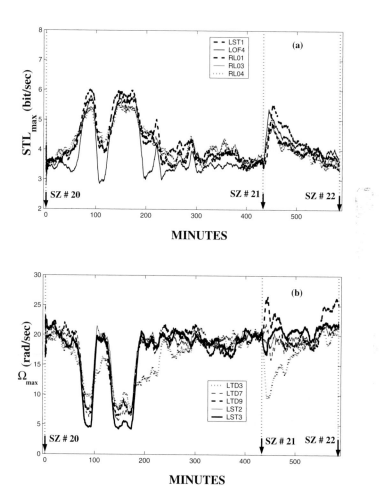

Figure 4.4. (a) Smoothed STL_{max} profiles of the 5 optimally selected electrodes over time (including seizures 20, 21, and 22). The optimal electrodes were selected 10 minutes before seizure 21. (b) Smoothed Ω_{max} profiles from the same EEG data as in (a).

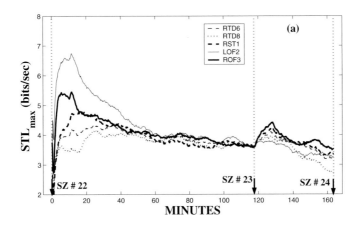

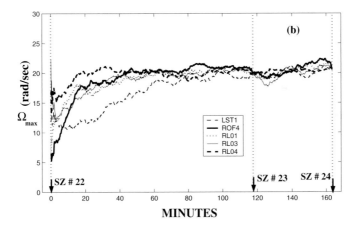

Figure 4.5. (a) Smoothed STL_{max} profiles of the 5 optimally selected electrodes over time (including seizures 22, 23, and 24). The optimal electrodes were selected 10 minutes before seizure 23. (b) Smoothed Ω_{max} profiles from the same EEG data as in (a).

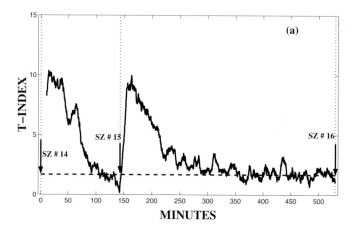

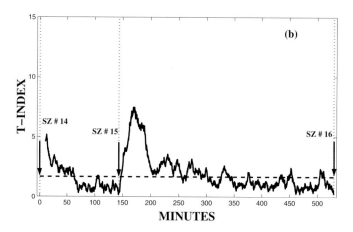

Figure 4.6.   (a) Average T-index curve over time from the STL_{max} profiles in Figure 3(a). (b) Average T-index curve over time from the Ω_{max} profiles in Figure 3(b).

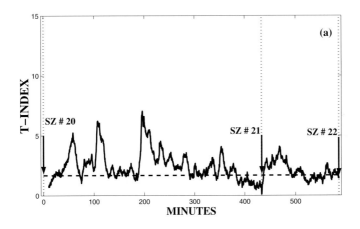

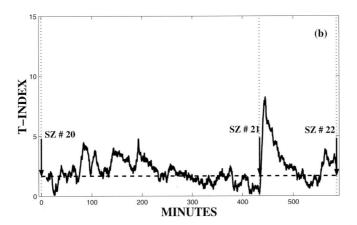

Figure 4.7. (a) Average T-index curve over time from the STL_{max} profiles in Figure 4(a). (b) Average T-index curve over time from the Ω_{max} profiles in Figure 4(b).

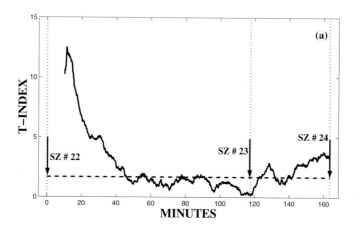

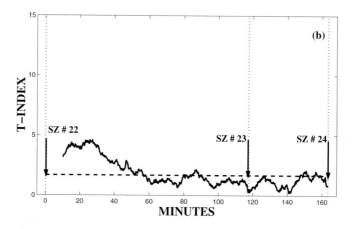

Figure 4.8. (a) Average T-index curve over time from the STL_{max} profiles in Figure 5(a). (b) Average T-index curve over time from the Ω_{max} profiles in Figure 5(b).

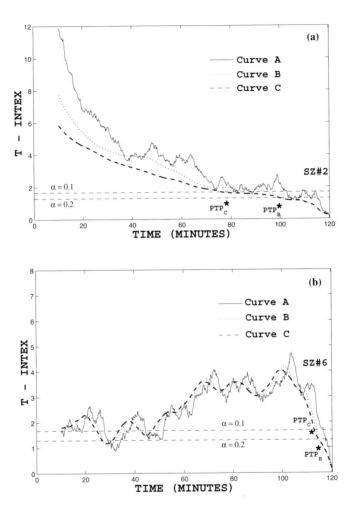

Figure 4.9. (a) An example of a predictable seizure by the average T-index curves of the preictally selected sites. Curve A: original T-index curve of the selected sites. Curves B and C: smoothed curves of A over windows of entrainment with length defined from critical values T_α at significance levels 0.2 and 0.1, respectively. (b) An example of an unpredictable seizure. The definitions of curves A, B, C are the same as in (a).

2 After estimation of PTP, the moving window length is set equal to PTP. We then estimate the average T-index values within each overlapping window of PTP length moving backwards in time up to 2 hours prior to a seizure's onset. A false positive is called if the average T-index within one of the windows is less than the critical value T_α that was used to determine the PTP.

3 Repeat steps 1 and 2 for $\alpha = \alpha'$.

4 A seizure is considered to be unpredictable if false positives are observed for both critical values α and α', at same or different time points, at most 2 hours prior to the seizure. Otherwise, the seizure is considered predictable.

Table 4.1. Predictability analysis by STL_{max} for 24 epileptic seizures in 1 patient. PTP_B and PTP_C are preictal transition periods for 2 significance levels. FP_B and FP_C are the number of the observed false positives by PTP_B and PTP_C, respectively.

SZ♯	PTP_B	FP_B	PTP_C	FP_C	Predictability (yes / no)
1	16.56	2	20.31	36	no
2	30.55	0	54.10	0	yes
3	14.34	49	24.75	0	yes
4	16.38	0	18.77	0	yes
5	43.35	0	98.47	0	yes
6	15.36	28	17.24	77	no
7	24.58	0	69.12	0	yes
8	61.78	18	120.00	0	yes
9	23.72	210	120.00	0	yes
10	76.47	0	90.96	0	yes
11	45.40	60	120.00	0	yes
12	24.06	0	28.33	14	yes
13	38.57	0	84.48	0	yes
14	20.65	0	40.79	0	yes
15	40.96	0	65.54	0	yes
16	120.00	0	120.00	0	yes
17	27.65	0	30.72	0	yes
18	120.00	0	120.00	0	yes
19	45.74	0	68.78	0	yes
20	18.09	0	18.09	0	yes
21	77.48	0	107.52	0	yes
22	15.70	234	20.65	376	no
23	88.24	0	100.18	0	yes
24	24.06	0	45.74	0	yes

Table 4.2. Predictability analysis by Ω_{max} for 24 epileptic seizures in 1 patient.

SZ♯	PTP_B	FP_B	PTP_C	FP_C	Predictability (yes / no)
1	30.21	202	120.00	0	yes
2	20.31	0	23.04	0	yes
3	45.40	0	67.41	0	yes
4	15.87	0	19.11	0	yes
5	20.48	29	28.33	109	no
6	16.04	5	18.26	43	no
7	69.80	0	82.60	0	yes
8	103.60	0	120.00	0	yes
9	33.62	0	53.93	0	yes
10	36.69	0	67.58	0	yes
11	37.03	295	120.00	0	yes
12	30.21	0	55.47	0	yes
13	66.05	0	91.82	0	yes
14	71.85	0	76.63	0	yes
15	120.00	0	120.00	0	yes
16	20.14	0	25.43	0	yes
17	120.00	0	120.00	0	yes
18	35.33	0	44.03	68	yes
19	41.81	0	58.20	0	yes
20	18.09	0	18.09	0	yes
21	120.00	0	120.00	0	yes
22	16.21	0	21.50	36	yes
23	76.63	0	98.30	0	yes
24	16.04	82	19.46	85	no

Table 4.3. Summary of predictability results for 24 epileptic seizures in 1 patient.

Preictal Transition Period (PTP) dervided from:	STL_{max}	Ω_{max}
PTP_B (minutes)	42.9 ± 6.5	49.2 ± 7.3
PTP_C (minutes)	66.9 ± 8.1	66.2 ± 8.1
Predictable seizures	21	21
Predictability	87.5%	87.5%

5.   Predictability results

In this section, results from the application of the previously described scheme to determine the predictability of epileptic seizures are shown. The method is applied to a patient with 24 seizures in 83.3 hours. The method described in the previous section was applied with two different critical values ($\alpha = 0.1, \alpha' = 0.2$).

Figures 9(a) and 9(b) show examples of a predictable and an unpredictable seizure, respectively. In both Figures, curves B ($\alpha = 0.2$) and C ($\alpha = 0.1$) are smoothed curves of A (by averaging the original T-index values within a moving window of length equal to PTP, which is different per curve, that is PTP_B for curve B and PTP_C for curve C). In Figure 9(a), the preictal transition period PTP_B identified by curve B is about 20 minutes, and PTP_C (identified by curve C) is about 43 minutes. It is clear that there are no false positives observed in both curves over the 2-hour period prior to this seizure. Thus, we conclude that this seizure is predictable. In Figure 9(b), the PTP's identified by the smoothed curves are 5 and 7 minutes, respectively. But false positives are observed prior to this seizure's onset (i.e. at 85 and 75 minutes for curves B and C, respectively). Therefore, we conclude that this seizure is not predictable.

Tables 1 and 2 show the results of the predictability analysis for each seizure by STL_{max} and Ω_{max}, respectively. Table 3 summarizes the predictability results of this analysis on the STL_{max} and Ω_{max} profiles for all 24 seizures of our patient. At first we observe that predictability is high by using STL_{max} or Ω_{max}. Second, we notice that the period of the preictal transition (PTP) is on the order of tens of minutes (ranges from 14 to 120 minutes) and varies according to the desired statistical significance level α. Third, Ω_{max}, the new measure used herein for quantification of the EEG dynamics, performs equally well to STL_{max} with respect to predictability of the seizures in this patient.

6.   Conclusions

This study suggests that it may be possible to predict focal-onset epileptic seizures by analysis of a new measure of the dynamics of EEG signals, namely the phase information of a state in the state space, recorded from multiple electrode sites. Previous studies by our group have shown that there is a preictal transition, in which the values of Lyapunov exponents of EEG recorded from critical electrode sites converge long prior to a seizure's onset. The electrode sites involved in this dynamical spatiotemporal interaction vary from seizure to seizure even in the same patient. Thus, the ability to predict a given seizure depends upon the ability to identify the critical electrode sites that participate in the preictal transition. The same conclusions are derived from the analysis of the new measure of dynamics proposed herein.

By employing a quadratic zero-one optimization technique for the selection of critical brain sites from the estimated spatiotemporal rate of change of phase (angular frequency) profiles, we demonstrated that 87.5% of the 24 seizures analyzed from 1 patient with right temporal lobe epilepsy were predictable. These results are comparable to those generated by the use of the STL_{max} profiles. Further studies are underway to compare the performance of STL_{max} and Ω_{max} measures on seizure predictability in a broader patient data base.

We believe that the proposed techniques may become valuable for on-line, real-time seizure prediction. Such techniques could also be incorporated into diagnostic and therapeutic devices for long-term monitoring and treatment of epilepsy.

References

[1] H. D. I. Abarbanel. *Analysis of observed chaotic data*. Springer-Verlag, New York, 1996.

[2] B. W. Abou-Khalil, G. J. Seigel, J. C. Sackellares, S. Gilman, R. Hichwa, and R. Marshall. Positron emission tomograghy studies of cerebral glucose metabolism in patients with chronic partial epilepsy. *Ann. Neurol.*, 22:480–486, 1987.

[3] H. Berger. Uber das elektroenkephalogramm des menchen. *Arch. Psychiatr. Nervenkr.*, 87:527–570, 1929.

[4] D. E. Burdette, Sakuraisy, T. R. Henry, D. A. Ross, P. B. Pennell, K. A. Frey, J. C. Sackellares, and R. Albin. Temporal lobe central benzodiazepine binding in unilateral mesial temporal lobe epilepsy. em Neurology, 45:934–941, 1995.

[5] R. Caton. The electric currents of the brain. *BMJ*, 2:278, 1875.

[6] C. Domb. In C. Domb and M. S. Green, editors, *Phase Transitions and Critical Phenomena*. Academic Press, New York, 1974.

[7] J. P. Eckmann, S. O. Kamphorst, D. Ruelle, and S. Ciliberto. Lyapunov exponents from time series. *Phys. Rev. A*, 34:4971–4972, 1986.

[8] C. E. Elger and K. Lehnertz. Seizure prediction by non-linear time series analysis of brain electrical activity. *Europ. J. Neurosci.*, 10:786–789, 1998.

[9] T. Elbert, W. J. Ray, J. Kowalik, J. E. Skinner, K. E. Graf, and N. Birbaumer. Chaos and physiology: Deterministic chaos in excitable cell assemblies. *Physiol. Rev.*, 74:1–47, 1994.

[10] J. Engel Jr., D. E. Kuhl, M. E. Phelps, and J. C. Mazziota. Interictal cerebral glucose metabolism in partial epilepsy and its relation to EEG changes. *Ann. Neurol.*, 12:510–517, 1982.

[11] M. A. Falconer, E. A. Serefetinides, and J. A. N. Corsellis. Aetiology and pathogenesis of temporal lobe epilepsy. *Arch. Neurol.*, 19:233–240, 1964.

[12] P. Gloor. *Hans Berger on the electroencephalogram of man*. Elsevier, Amsterdam, 1969.

[13] P. Grassberger, T. Schreiber, and C. Schaffrath. Nonlinear time sequence analysis. *Int. J. Bifurc. Chaos*, 1:521–547, 1991.

[14] A. V. Holden. *Chaos-nonlinear science: theory and applications*. Manchester University Press, Manchester, 1986.

[15] H. Horst, P. M. Pardalos , and V. Thoai. *Introduction to global optimization, Series on Nonconvex Optimization and its Applications, 3*. Kluwer Academic Publishers, Dordrecht, 1995.

[16] L. D. Iasemidis, J. C. Sackellares, H. P. Zaveri, and W. J. Williams. Phase space topography of the electrocorticogram and the Lyapunov exponent in partial seizures. *Brain Topogr.*, 2:187–201, 1990.

[17] L. D. Iasemidis. *On the dynamics of the human brain in temporal lobe epilepsy*. Ph.D. thesis, University of Michigan, Ann Arbor, 1991.

[18] L. D. Iasemidis and J. C. Sackellares. The temporal evolution of the largest Lyapunov exponent on the human epileptic cortex. In D. W. Duke and W. S. Pritchard, editors, *Measuring chaos in the human brain*. World Scientific, Singapore, 1991.

[19] L. D. Iasemidis, J. C. Principe, and J. C. Sackellares. Spatiotemporal dynamics of human epileptic seizures. In R. G. Harrison, W. Lu, W. Ditto, L. Pecora, M. Spano, and S. Vohra, editors, *3rd Experimental Chaos Conference*. World Scientific, Singapore, 1996.

[20] L. D. Iasemidis and J. C. Sackellares. Chaos theory and epilepsy. *The Neuroscientist*, 2:118–126, 1996.

[21] L. D. Iasemidis, J. C. Principe, J. M. Czaplewski, R. L. Gilman, S. N. Roper, and J. C. Sackellares. Spatiotemporal transition to epileptic seizures: A nonlinear dynamical analysis of scalp and intracranial EEG recordings. In F. Lopes da Silva, J. C. Principe, and L. B. Almeida, editors, *Spatiotemporal Models in Biological and Artifical Systems*. IOS Press, Amsterdam, 1997.

[22] L. D. Iasemidis, D. S. Shiau, J. C. Sackellares, and P. M. Pardalos. Transition to epileptic seizures: Optimization. In D. Z. Du, P. M. Pardalos and J. Wang, editors, *DIMACS series in Discrete Mathematics and Theoretical Computer Science, vol. 55*. American Mathematical Society, 1999.

[23] L. D. Iasemidis, P. M. Pardalos, J. C. Sackellares, and D. S. Shiau. Quadratic binary programming and dynamical system approach to determine the predictability of epileptic seizures. *Journal of Combinatorial Optimization.* 5:9-26, 2000.

[24] L. D. Iasemidis, P. M. Pardalos, J. C. Sackellares, and D. S. Shiau. Global optimization and nonlinear dynamics to investigate complex dynamical transitions: Application to human epilepsy. *IEEE Transactions on Biomedical Engineering.* in press.

[25] B. H. Jansen. Is it and so what? A critical review of EEG-chaos. In D. W. Duke and W. S. Pritchard, editors, *Measuring chaos in the human brain.* World Scientific, Singapore, 1991.

[26] A. N. Kolmogorov. The general theory of dynamical systems and classical mechanics. In R. Abraham and J. E. Marsden, editors, *Foundations of Mechanics.* 1954.

[27] E. J. Kostelich. Problems in estimating dynamics from data. *Physica D*, 58:138–152, 1992.

[28] K. Lehnertz, and C. E. Elger. Can epileptic seizures be predicted? Evidence from nonlinear time series analysis of brain electrical activity. *Phys. Rev. Lett.*, 80:5019–5022, 1998.

[29] F. Lopes da Silva. EEG analysis: theory and practice; Computer-assisted EEG diagnosis: Pattern recognition techniques. In E. Niedermeyer and F. Lopes da Silva, editors, *Electroencephalography: Basic principles, clinical applications and related field.* Urban and Schwarzenberg, Baltimore, 1987.

[30] M. Le Van Quyen, J. Martinerie, M. Baulac, and F. Varela. Anticipating epileptic seizures in real time by a non-linear analysis of similarity between EEG recordings. *NeuroReport*, 10:2149–2155, 1999.

[31] J. Martinerie, C. Adam, M. Le Van Quyen, M. Baulac, S. Clemenceau, B. Renault, and F. J. Varela. Epileptic seizures can be anticipated by non-linear analysis. *Nature Medicine*, 4:1173–1176, 1998.

[32] J. H. Margerison and J. A. N. Corsellis. Epilepsy and the temporal lobes. *Brain*, 89:499–530, 1966.

[33] G. Mayer-Kress. *Dimension and entropies in chaotic systems.* Springer-Verlag, Berlin, 1986.

[34] J. W. McDonald, E. A. Garofalo, T. Hood, J. C. Sackellares, S. Gilman, P. E. McKeever, J. C. Troncaso and M. V. Johnston. Altered excitatory

and inhibitory aminoacid receptor binding in hippocampus of patients with temporal lobe epilepsy. *Annals of Neurology*, 29:529–541, 1991.

[35] A. Oseledec. A multiplicative ergodic theorum-Lyapunov characteristic numbers for dynamical systems (English translation). *IEEE Int. Conf. ASSP*, 19:179–210, 1968.

[36] N. H. Packard, J. P. Crutchfield, J. D. Farmer, and R. S. Shaw. Geometry from time series. *Phys. Rev. Lett.*, 45:712–716, 1980.

[37] J. Pesin. Characteristic Lyapunov exponents and smooth ergodic theory. *Russian Math. Survey*, 4:55–114, 1977.

[38] J. C. Sackellares, G. J. Siegel, B. W. Abou-Khalil, T. W. Hood, S. Gilman, P. McKeever, R. D. Hichwa, and G. D. Hutchins. Differences between lateral and mesial temporal metabolism interictally in epilepsy of mesial temporal origin. *Neurology*, 40:1420–1426, 1990.

[39] D. L. Stein. In D. L. Stein, editor, *Lecture Notes in the Sciences of Complexity, SFI Studies in the Science of Complexity*. Addison-Wesley Publishing Company, 1989.

[40] F. Takens. Detecting strange attractors in turbulence. In D. A. Rand and L. S. Young, editors, *Dynamical systems and turbulence, Lecture notes in mathematics*. Springer-Verlag, Heidelburg, 1981.

[41] J. A. Vastano and E. J. Kostelich. Comparison of algorithms for determining Lyapunov exponents from experimental data. In G. Mayer-Kress, editor, *Dimensions and entropies in chaotic systems: quantification of complex behavior*. Springer-Verlag, Berlin, 1986.

[42] P. Walters. *An intorduction to ergodic theory*. Springer-Verlag, Berlin, 1982.

[43] A. Wolf, J. B. Swift, H. L. Swinney, and J. A. Vastano. Determining Lyapunov exponents from a time series. *Physica D*, 16:285–317, 1985.

Chapter 5

SELF-ORGANIZING MAPS
Multivariate Learning Algorithms and Applications to Auditory Spike Train Analysis

Jennie Si
Department of Electrical Engineering
Arizona State University
si@asu.edu

Daryl R. Kipke
Department of Bioengineering
Arizona State University
kipke@asu.edu

Russell Witte
Department of Bioengineering
Arizona State University
nails@asu.edu

Jing Lan
Department of Electrical Engineering
Arizona State University
Jing.Lan@asu.edu

Siming Lin
DSP-Vision Group
National Instruments
siming.lin@ni.com

P.M. Pardalos and J. Principe (eds.), Biocomputing, 85-106.

Abstract   Self-organizing map (SOM) has been applied in many different fields of science
and engineering. In this chapter it is shown how the SOM can be used to decode
neural spike trains of an awake animal and associate external auditory stimuli
with spike patterns of the brain. This chapter begins with an introduction of the
SOM and highlights advantages of the SOM when compared to other multivariate
statistical data analysis tools commonly used to analyze similar data sets. In this
study, simultaneous multichannel recording from guinea pig auditory cortex is
examined using the SOM to assess the effectiveness and potential of the SOM in
neurophysiological studies.

Keywords: Self-organizing map, Multivariate analysis, Neuronal ensemble decoding, Cortical recording.

Introduction

Previous neurophysiologic experiments have identified two major topograph-
ically organized computational maps in primary visual cortex of cats and monkeys [1] [2] [3] [4]:

1 Maps of preferred line orientation, representing the angle of tilt of a line
   stimulus.

2 Maps of ocular dominance, representing the relative strengths of excitatory influence of each eye.

Electrophysiologic studies have further demonstrated that other sensory processing areas of the brain, including the cerebral cortex, are spatially organized
according to input stimuli [4][5]. For instance, tonotopic maps of the auditory
cortex indicate that different sound frequencies have a predictable representation based on the anatomical position in the cortex (e.g., in the guinea pig, lower
frequencies are more represented anterior.) [4] [5].

The idea of SOM in artificial neural networks may be traced back to the
early work of von der Masburg [6] [7] in the 1970s on the self-organization of
orientation sensitive nerve cells in the striate cortex. In 1976, Willshaw and von
der Masburg [7] published the first paper on the formation of self-organizing
maps on biological grounds to explain retinotopic mapping from the retina to
the visual cortex (in higher vertebrates). But it was not until the publication of
Kohonen's paper on the self-organizing feature map [8] in 1982, that the SOM
emerged as an attractive tool to model biological systems. Kohonen's model
uses a computational shortcut to mimic basic functions similar to biological
neural networks. The SOM was used to create an "ordered" map of input
signals based on 1) the internal structure of the input signals themselves and 2)
the coordination of the unit activities through the lateral connections between
the units. Many implementation details of biological systems were "ignored" in
Kohonen's SOM model. Instead, the SOM was an attempt to develop a model

based on heuristically conceived, although biologically inspired, functional structure.

The SOM learning algorithm is a statistical data modeling tool which has two distinct properties: 1) Clustering of multi-dimensional input data; and 2) Spatially ordering the output map so that similar input patterns tend to produce a response in units that are close to each other in the output map. While performing the statistical data analysis function, the SOM is also a convenient tool for visualizing results from multi-dimensional data analysis.

Today, the SOM has found widespread applications in various areas, such as robotics [9] [10] [11], process control [12] [13] [14], speech processing [15] [16] [17] , library and document management [18] [19], image analysis and pattern recognition [20] [21] [22], power system control [23], multisource data fusion [24] [25], biological system analysis [26] [27] [28], semiconductor manufacturing [22], computer systems [29], and health care systems [30] [31].

This chapter uses the SOM technique to decode neural spike trains recorded from a multichannel electrode array implanted in a guinea pig's auditory cortex. After introducing the SOM, the technique will be compared to other popular multivariate data analysis algorithms. The remainder of the chapter will explicitly focus on applying the SOM to simultaneous multichannel recording - from preparing the data, to selecting the appropriate algorithm parameters. The effectiveness of the SOM algorithm will be illustrated by comparing the results of frequency discrimination generated by SOM and statistical template matching.

1. The Self-Organizing Map Algorithm

To begin with the introduction of SOM, the concept of an "undirected graph" is needed. An undirected graph G is a pair $G = (R_g, E_g)$, where $R_g = \{r_1, r_2, \cdots, r_L\}$ is a finite set of nodes or vertices in an m dimensional space with $r_i \in \mathcal{R}^m$, $i = 1, 2, \cdots, L$. A node in G can be represented by its index i or its position r_i. E_g is a subset of the undirected edges, $E_g \subset \{[r_i, r_j] \mid r_i, r_j \in R_g$ and $i \neq j\}$. If $e_{ij} = [r_i, r_j] \in E_g$, then we say that node r_i is adjacent to node r_j (and vice versa) in G.

The SOM is intended to learn the topological mapping $f : \mathcal{X} \subset \Re^n \rightarrow G \subset \Re^m$ by means of self-organization driven by samples X in $\mathcal{X}$, where G is a graph which determines the output map structure containing a set of nodes, each representing an element in the m-dimensional Euclidean space. Let $X = [x_1, x_2, \cdots, x_n]^T \in \mathcal{X}$ be the input vector. It is assumed to be connected in parallel to every node in the output map. The weight vector of node i is denoted by $W_i = [w_{i1}, w_{i2}, \cdots, w_{in}]^T \in \Re^n$ (see Figure 5.1).

In Kohonen's SOM algorithm, the graph (output map) is usually prespecified as a one-dimensional chain or two dimensional lattice. A node in G can be represented by its index i or its position r_i. In this chapter, we will assume that only one-dimensional chain with L nodes or two dimensional lattice with $L = L_x \times L_y$ nodes is used as an output map. Therefore, if the chain is used as the output map, we let $r_i = i, i = 1, 2, \cdots, L$. If the lattice is used as an output map, we let $r_i = (i_x, i_y)$.

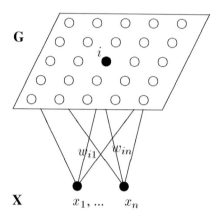

Figure 5.1.   The self-organizing feature mapping architecture.

1.1.   SOM building blocks

There are two important steps in the self-organization algorithm:

1 Selecting neighborhood: find the activity bubble.

2 Adaptive process: update the weight vectors using the Hebbian learning law.

A computationally efficient procedure to accommodate the two steps can be elaborated as follows.

Find the winner and the neighborhood of the winner: Find the output node with the largest response to the input vector. This can be done by simply comparing the inner products $W_i^T X$ for $i = 1, 2, \cdots, L$ and selecting the node with the largest inner product. This process has the same effect in determining the location where the activity bubble is to be formed. If the weight vectors W_i are normalized, the inner product criterion is equivalent to the minimum Euclidean distance measure. Specially, if we use the index $c(X)$ to indicate the output node of which the weight vector "matches" the input vector X the best, we may then determine $c(X)$ by applying the following condition.

$$c(X) = \arg \min_i \|X - W_i\|, \qquad i = 1, 2, \cdots, L \qquad (1)$$

where $\| \cdot \|$ denotes the Euclidean norm of the argument. In this chapter, we use Euclidean norm as the metric unless a specification is made. In most cases, the Euclidean metric can be replaced by others with different characteristics in mind.

The activity bubble is implemented by a topological neighborhood N_c of the node c. The neighborhood N_c depends on the output map structure used in the SOM.

Adaptive process: The weight vectors inside the neighborhood of the winner are usually updated by Hebbian type learning law. The Oja's rule states that [32]:

$$\frac{dW_i(t)}{dt} = \alpha(t)y_i(t)\{X(t) - y_i(t)W_i(t)\}, \qquad i = 1, 2, \cdots, L \qquad (2)$$

where the negative component is a nonlinear *forgetting term*. We may further simplify Equation (2) by assigning the output y_i a binary value. $y_i(t) = 1$ if the node i is inside the activity bubble, i.e., inside the neighborhood of the winner c; $y_i(t) = 0$ otherwise. Therefore, we can rewrite equation (2) as follows:

$$\frac{dW_i(t)}{dt} = \begin{cases} \alpha(t)(X(t) - W_i(t)) & \text{for } i \in N_c \\ 0 & \text{otherwise.} \end{cases} \qquad (3)$$

The most convenient and widely accepted SOM algorithm was obtained in discrete-time format by Kohonen as follows [8] [33],

$$W_i(k+1) = \begin{cases} W_i(k) + \alpha(k)(X(k) - W_i(k)) & \text{for } i \in N_c \\ W_i(k) & \text{otherwise.} \end{cases} \qquad (4)$$

1.2. Implementation of the SOM algorithm

The essence of the Kohonen's SOM algorithm is to take a computational "shortcut" to achieve the effect accomplished by the typical "Mexican hat" lateral interactions. What follows is a step by step learning procedure of Kohonen's SOM algorithm.

step 1 Initialize the weights by random numbers,

$$W_i(0) = (w_{i1}(0), w_{i2}(0), \cdots, w_{in}(0)) \in \Re^n, i = 1, 2, \cdots, L.$$

step 2 Draw a sample X from the input distribution.

step 3 Find the winner W_c (best-matching node) using the minimum Euclidean distance criterion:

$$c(X) = \arg \min_i \|X(k) - W_i\|, \qquad i = 1, 2, \cdots, L. \qquad (5)$$

step 4 Update the winner and its neighbors by

$$W_i(k+1) = W_i(k) + \alpha(k)\Lambda(i, c)[X(k) - W_i(k)], \quad (6)$$

where $k = 0, 1, \cdots$ denotes the discrete time steps, $\alpha(k)$ is the learning rate and $\Lambda(i, c)$ is the neighborhood function of the winner.

Step 5 Compute $E_k = \sum_i \|W_i(k+1) - W_i(k)\|$, if $E_k \leq \epsilon$ stop; else repeat from step 1.

In the above, $\Lambda(i, c)$ is a neighborhood function. $\Lambda(i, c)$ equals one for $i = c$ and falls off with the distance $\|r_c - r_i\|$ between node i and the winner c in the output layer, where r_c and r_i denote the coordinate positions of the winner c and node i in the output layer, respectively. Thus, those nodes close to the winner, as well as the winner c itself, will have their weights changed appreciably, while those farther away, where $\Lambda(i, c)$ is small, will experience little effect. It is here that the topological information is supplied: nearby nodes receive similar updates and thus end up responding to nearby input patterns. In the original learning algorithm proposed by Kohonen [34], the neighborhood function $\Lambda(i, c)$ is defined as

$$\Lambda(i, c) = \begin{cases} 1 & \text{for } \|r_i - r_c\| \leq N_c(k) \\ 0 & \text{otherwise} \end{cases} \quad (7)$$

where $N_c(k)$ is some decreasing function of time. The value of $N_c(k)$ is usually large at the beginning of learning and then shrinks during training. We call this neighborhood function a square neighborhood function.

The bell-shaped neighborhood function [16] is also frequently used in practice:

$$\Lambda(i, c) = exp(-\|r_i - r_c\|^2 / 2\sigma^2(k)) \quad (8)$$

where $\sigma(k)$ is the width parameter that affects the topology order in the output map and is gradually decreasing during training.

The learning rate $\alpha(k)$ in the learning algorithm is essential for convergence. The learning rate $\alpha(k)$ should be large enough so that the network could adapt quickly for the new training patterns. On the other hand, $\alpha(k)$ should be small enough so that the network would not forget the experience from the past training patterns. For analytical purposes, $\alpha(k)$ could be chosen to satisfy conditions in Robbins-Monro algorithm [35] [36]. In the update scheme shown above, the winner and its neighborhood is simply leaning toward the current input pattern by moving along the vector $(X(k) - W_i)$ that pushes W_i toward X. The amount of adjustment of W_i depends on the value of the learning rate parameter $\alpha(k)$, which varies from 0 to 1. If $\alpha(k) = 0$, there is no update; and when $\alpha(k) = 1$, W_c becomes X.

For effective global ordering of the output map, it has been observed experimentally to be advantageous to let the width parameter σ be very large in the beginning and gradually decrease during the learning process. As a matter of fact it was shown in [36] that if the range of the neighborhood function covers the entire output map, then each weight vector converges to the same stationary state, which is the mass center of the training data set. This implies that if we want to eliminate the effect of the initial conditions, we should use a neighborhood function covering a large range of the output map. On the other hand, if the range of the neighborhood function becomes 0, that is, $N_c(k) = 0$, the final iterations of the SOM algorithm may be viewed as a sequential updating process of vector quantization. It is interesting to note that even though the convergence arguments are made based on Robbins-Monro algorithm, there has been no claims about topological ordering of the weight vectors. It remains as a well-observed practice in many applications.

2. Related Statistical Algorithms: A Qualitative Comparison

The SOM performs two important tasks: multidimensional clustering and preservation of the topology of the input space. In this section, we will discuss some statistical algorithms that are related to the SOM.

2.1. Vector quantization

Vector quantization is a technique that exploits the underlying structure of input vectors for the purpose of data compression. Specifically, an input space is divided into a number of distinct regions, and for each region a reconstruction vector is defined. A vector quantizer can be formally defined as a mapping Q from the n-dimensional Euclidean space $\mathcal{R}^n$ into a finite subset of $\mathcal{R}^n$: $\hat{A} = \{Y_i \in \mathcal{R}^n; i = 1, \cdots, L\}$. Thus, $Q : \mathcal{R}^n \to \hat{A}$.

Given an input vector space, the quantizer is specified by the values of the representative vectors $\hat{A} = \{Y_i \in \mathcal{R}^n; i = 1, \cdots, L\}$ and by the associated partition $P(\hat{A}) = \{V_i; i = 1, \cdots, L\}$. For an optimal VQ quantizer, the partition $P(\hat{A})$ associated with the set $\hat{A}$ of representative vectors is a Voronoi partition of the input space, i.e, V_i is the Voronoi polyhedron.

Let $X = (x_1, x_2, \cdots, x_n) \in \mathcal{R}^n$ be an n-dimensional input vector, then the L-level quantizer can be expressed as

$$Q(X) = Y_i, \quad \text{if } X \in V_i, \quad \text{for } i = 1, 2, \cdots, L. \tag{9}$$

Let $p(X) = p(x_1, x_2, \cdots, x_n)$ be a joint probability density function of X, then with the fixed rate (i.e., a fixed n and L), the average distortion is defined as follows:

$$D(Q) = \frac{1}{L}E[\|X - Q(X)\|^2] = \frac{1}{L}\sum_{i=1}^{L}\int_{V_i}\|X - Y_i\|^2 p(X)dX. \qquad (10)$$

To obtain an optimal vector quantizer, the distortion D(Q) should be minimized with respect to the partition $P(\hat{A})$ and the representative vectors $\hat{A} = \{Y_i \in \mathcal{R}^n; i = 1, \cdots, L\}$. The two conditions necessary for minimizing D in equation (10) are known as [37]:

Voronoi partition: Given representative vectors, the partition must be the Voronoi partition. That is, for fixed representative vectors, the optimal partition should be constructed in such a manner that:

$$Q(X) = Y_i, \qquad \text{iff } \|X - Y_i\| \leq \|X - Y_j\|, \quad \text{for all } j. \qquad (11)$$

Generalized centroid assignment: Given a partition, the representative vectors must be the generalized centroids. That is, for a fixed partition, the optimal representative vector Y_i should be computed so that:

$$E[\|X - Y_i\|^2 \mid X \in V_i] = \inf_{u \in \mathcal{R}^n} E[\|X - u\|^2 \mid X \in V_i], \qquad (12)$$

where $E[\cdot]$ is the expectation.

Note that the above two conditions are sufficient for a local minimum of D, but are only necessary for a global minimum. The first VQ design algorithm using the above conditions as a design guideline is the LBG algorithm [37].

It is interesting to note that the SOM can be regarded as an adaptive (or online) version of the LBG algorithm [36] when the range of the neighborhood function in SOM shrinks to 0 in finite steps. By saying so, we are referring to the clustering ability of the SOM. The LBG or vector quantization algorithm in general does not preserve the topology of the input data.

2.2.   Sammon's mapping

Sammon's mapping was proposed by Sammon [38] for nonlinear data projection from a high dimensional space down to two-dimensional. Sammon's mapping attempts to directly approximate local geometric relations among input samples in a two-dimensional space while preserving all the inter-pattern distances in input space. Let $X_i, i = 1, 2, \cdots, P$, be the P n-dimensional input patterns, and $\hat{X}_i = (\hat{X}_{i1}, \hat{X}_{i2}), i = 1, 2, \cdots, P$ be the P corresponding patterns in the two-dimensional projected space. Sammon's mapping is a direct result to minimize the following mapping error, also known as Sammon's stress :

$$E = \frac{1}{\sum_{i=1}^{P-1}\sum_{j=i+1}^{P}d(X_i, X_j)}\sum_{i=1}^{P-1}\sum_{j=i+1}^{P}\frac{d(X_i, X_j) - \hat{d}(\hat{X}_i, \hat{X}_j)}{d(X_i, X_j)} \qquad (13)$$

where $d(X_i, X_j)$ is the distances between patterns X_i and X_j in the input space, and $\hat{d}(\hat{X}_i, \hat{X}_j)$ is the distances between patterns $\hat{X}_i$ and $\hat{X}_j$ in the output space.

Sammon's stress E is a measure of how well inter-pattern distances are preserved when the patterns are projected from a high dimensional space down to two-dimensional.

To find the P patterns in the two-dimensional space that minimize E, Sammon used a gradient descent iterative process. This approach views the projected mapping positions as $2 \times P$ variables in the optimization problem. The following $2 \times P$ equations represent the iterative process of solving the $2 \times P$ position variables using gradient descent.

$$\hat{X}_{ip}(k+1) = \hat{X}_{ip}(k) - \alpha \frac{\frac{\partial E(k)}{\partial \hat{X}_{ip}(k)}}{\left| \frac{\partial^2 E(k)}{\partial^2 \hat{X}_{ip}(k)} \right|} \qquad i = 1, 2, \cdots, P, \qquad p = 1, 2. \quad (14)$$

Sammon's mapping is a nonlinear data projection algorithm just as the SOM. It is useful for preliminary analysis in statistical pattern recognition. It can be used to visualize class distributions to some extent, especially the degree of overlap between classes. Sammon's mapping projects P patterns from a higher-dimensional space down to two-dimensional. In contrast, the SOM can project the input data to a prespecified number of output nodes, which is usually less than P by choice. Sammon's algorithm involves significant amount of computation. For each iteration during the iterative gradient descent process, $n(n-1)/2$ computation of the distances is first needed and $2 \times P$ updating equations are then to be solved. For a large number of input patterns, Sammon's algorithm becomes impractical.

Compared with the SOM, Sammon's mapping has another disadvantage. It does not generalize. To project new data, one has to run the program all over again to include both the old and the new data. Mao [39] proposed a three-layer feedforward neural network model to realize Sammon's mapping with generalization capability, but the computational complexity is still high.

2.3. Principal Component Analysis

Principal component analysis (PCA) is a well known algorithm for data analysis. It is a linear orthogonal transform from an n-dimensional input space $\mathcal{R}^n$ to an m-dimensional space $\mathcal{R}^m$ with $m < n$, and it retains the maximal variance in the input data during the transform. Without loss of generality, assume the input $X \in \mathcal{R}^n$ has zero mean. Let $R = E[XX^T]$ be the covariance matrix. Let the eigenvalues of R be denoted by $\lambda_1, \lambda_2, \cdots, \lambda_n$, and the associated eigenvectors be denoted by $u_1, u_2, \cdots, u_n$, respectively. We may then write

$$Ru_j = \lambda_j u_j \qquad j = 1, 2, \cdots, n. \quad (15)$$

Let the first m eigenvalues be arranged in descending order:

$$\lambda_1 > \lambda_2 > \cdots > \lambda_m.$$

Let the associated m eigenvectors be used to construct an n-by-m matrix:

$$\Phi = [u_1, u_2, \cdots, u_m].$$

The matrix Φ is an orthogonal matrix in the sense that its column vectors are orthonormal to each other. The m principal components of the input data vector $X \in \mathcal{R}^n$ can be obtained by a linear transform

$$Y = \Phi^T X$$

where Y is the output vector in the m-dimensional space. The variance of the original data retained in the new m-dimensional space is $\sum_{i=1}^m \lambda_i$, which is the largest value among all linear orthogonal transforms of the same output dimensionality. PCA is optimal in the mean-square error sense among all linear orthogonal transforms [40]. PCA can be realized by artificial neural networks in various ways. A PCA network is a one-layer feedforward neural network [41] that is able to extract the principal components from the stream of input vectors. Typically Hebbian learning rules are used.

PCA can be used for multivariate data projection. For example, PCA can be used to project a multi-dimensional input data set into a two-dimensional space. In this case, the PCA map is spanned by two orthonormal vectors (the first two eigenvectors) along which the data have the largest and the second largest variances. PCA is a convenient tool to see major data distribution trend. For some specific input data structure, the PCA map may also exhibit most of the class-discriminatory information [39]. However, the PCA map is a linear mapping. And it is limited for that reason.

Compared with the PCA mapping, the SOM has the following advantages: 1) the SOM implements nonlinear mapping; thus, it is better to use the SOM with highly convolved input data. 2) the SOM attempts to preserve the neighborhood relations between the input space and output map as much as possible. It can be used to exhibit details of the input data structure in the output map. Theoretically, the PCA is optimal for data compression. It sacrifices classification accuracy for nonlinearly related data sets.

There are other related algorithms to the SOM, such as linear discriminant analysis (LDA), non-linear discriminant analysis (NDA), independent component analysis (ICA) and others. Although these algorithms may be used for multivariate data projection, they are often difficult to implement with on-line learning [39] or computationally expensive. These algorithms are mainly for special purpose applications. For example, ICA has been used for source separation problems [42].

3.   Background of Decoding Auditory Recordings

A simple tone pip played in the external environment elicits a complex cascade of events. They ultimately lead to our perception of sound. Between the one-dimensional representation of frequency along the cochlea to the two-dimensional tonotopic sheet of neurons in the auditory cortex, the central auditory system parcellates the original stimulus into multiple anatomical fields. The original stimulus undergoes massive diffusion, as the coding of the sound diverges from the auditory nerve, up the brain stem, and to the cortex. Conversely, comparatively fast firing rates in the auditory nerve eventually become relatively slow firing neurons in the cortex.

Such complex patterns of spatial diffusion and temporal parcellation of the sound stimulus, however, must eventually converge to a particular behavior, whether that be discriminating two frequencies or understanding the spoken language. In this section, we examine multichannel cortical recordings from a guinea pig, and associate the recorded cortical activities with external sound stimuli, which are represented as tone pips at different frequencies.

3.1.   Simultaneous cortical recordings

A guinea pig was implanted with a chronic array of 33 microelectrodes (3 rows of 11 microwires each) in auditory cortex. Simultaneous neural recordings were obtained from the awake guinea pig as soon as the day after the surgical implant. Each wire measured the extracellular potential in the cortex. Signals were buffered using JFET amplifiers and then sent to a multichannel neural recording system (Plexon, Inc., Dallas, TX). The system offered differential amplification, 40 kHz sampling, and band pass filtering (50-12000 Hz). Online spike discrimination allowed for unit separation and identification using template matching and/or PCA. Recorded neural signals were typically between 1 and 4 neurons, depending on the discriminability of the waveforms. Both timestamps and waveform data could be saved to a hard disk. The guinea pig in this study was restrained in a sling, and the head was not immobilized.

Stimulus generation and behavioral feedback were controlled by a computer-based sound system (Tucker-Davis Technologies Inc.). Each auditory stimulus was delivered within an enclosed anaechoic sound chamber in the free field using a single high-performance speaker placed 18 inches above the animal's head. Frequency response areas (FRAs) were used to map unit responses to pure tone stimuli [43] and were routinely presented to describe receptive fields of recorded auditory neurons. The FRA stimulus used to predict frequency classes consisted of a pseudorandom sequence of 60 frequencies at a single sound intensity. In the example provided in this study, each frequency was repeated ten times. The total stimulation time for an individual FRA was approximately 10 minutes.

3.2.   Neuron firing rate calculations from recorded spike trains

While the guinea pig sat in the sound chamber listening passively to the tone pips of different frequency, simultaneous recording of action potentials (spikes) were detected in the extracellular space generated by auditory cortical cells near the tip of each electrode (within about 50 microns). The result of these recordings is a spike train at the following typical time instances, $\tau_1, \tau_2, \cdots, \tau_N$. This spike train can be ideally modeled by a series of δ functions at the appropriate times:

$$S_c(t) = \sum_{i=1}^{N} \delta(t - \tau_i). \tag{16}$$

Let Ω be the collection of the guinea pig's auditory cortex cells whose activities were stimulated by the external tone pips. Realistically we can only record a set of sample cells $S = \{S_i \in \Omega, i = 1, 2, \cdots, n\}$ to analyze the relation between discharge rate patterns in the guinea pig's auditory cortex and the external sound stimuli. The recording procedure assumes that the distribution of sample cells uniformaly covers the auditory map in the range of the stimulus frequencies. The firing activity of each cell S_i defines a point process $f_{s_i}(t)$. Let d_{ij} denote the average discharge rate for the ith cell at the jth interval $(t_j, t_j + dt)$. Taking d_{ij} of all the cells $\{S_i \in \Omega, i = 1, 2, \cdots, n\}$ in $(t_j, t_j + dt)$, we obtain a vector $X_j = [d_{1j}, d_{2j}, \cdots, d_{nj}]$ which will later be used as input vectors relating to the external stimulus through various statistical analysis including SOM clustering.

The average discharge rate $d_{ij} = \frac{D_{ij}}{dt}$ is calculated from the raw spike train as follows. For the jth time interval $(t_j, t_j + dt)$, one has to first determine all the intervals between any two neighboring spikes which overlap with the jth time interval. Each of the spike intervals overlapping completely with $(t_j, t_j + dt)$ contributes a 1 to D_{ij}. Each of those only overlapping partially with $(t_j, t_j + dt)$ will contribute the percentage of the overlapping to D_{ij}.

In our analysis, spike signals from 33 electrodes in the auditory cortex were used. Each electrode usually detects from one to four neuronal signals depending on the discriminability of the waveforms. Various levels of spatial ensemble average and temporal ensemble average were used in preparing the firing rates before a clustering analysis was performed.

3.3.   Preliminary analysis using template matching

In the experiment that is to be analyzed, 33 electrodes were implanted in the auditory cortex of a guinea pig. That has resulted in about 70 discriminated waveforms in the original date file. The files recorded the spike time instances over a course of one complete trial. One trial (defined as one stimulus interval) lasted 700ms, including 200 ms of tone on time (with 5 ms rise/fall) and 500 ms of off time (interstimulus interval).

The spike patterns entail two types of randomness. First, each neuron spikes randomly for each trial for the same stimulus, although statistically it has been approximately modeled by a Poisson process [44]. Second, a spike pattern in a neural ensemble may vary for the same stimulus, possible as a consequence of redundancy in the neural code. These combined factors make it a challenge to decode neural ensemble spike patterns and relate them to the acoustic stimuli. As a result, spatial and temporal averages are often used to reduce the dimensionality of the firing patterns.

Figure 5.2 is an example of spatial averages taken in order to discriminate two firing patterns associated with two stimulus frequencies. Although spatial and temporal averaging may be useful for visualizing the firing patterns, they have limited value when it comes to real-time recognition of stimulus frequency without the benefit of abundant repetitions. Figure 5.2 illustrates different types of temporal coding patterns of the signal with clear onset, sustained, and offset response types shown. In many cases, however, the distinction between the responses are not as great as shown in Figure 5.2.

To quantitatively evaluate the level of similarity/dissimilarity between the waveforms corresponding to different frequency stimuli, we performed a correlation analysis between the 60 waveforms (two of which are shown in Figure 5.2). The template waveforms for the 60 frequencies were created using 9 out of the 10 trials used throughout this analysis. And the one trial left out was used for testing the class membership. We have used the *Leave-One-Out* procedure in this analysis. If the waveforms as shown in the bottom row of Figure 5.2 are used directly for this template matching analysis, only 3% of the testing patterns were correctly identified, i.e., we can only associate successfully 3% of the spike patterns with the external stimuli. An improvement of up to 11% successful recognition rate was obtained when a Gaussian filter is applied to each of the waveforms. This preliminary analysis reveals that simple statistical analysis is far from satisfactory when dealing with complex signals that are correlated both spatially and temporally.

4. Relating the spike Pattern of Auditory Neurons to the Sound Stimuli using SOM

In the application considered herein, the SOM is used as a decoder. By finding the winners, the SOM algorithm maps an input (a cortical spike pattern) onto a node in the output layer (a stimulus frequency). Or in other words, the association between spike patterns and external stimulus frequencies are made through the weight vectors in the SOM.

In our current application, we have used a two-dimensional lattice as the output map in SOM. The sizes of the map were tuned according to training data to find a best fitting schedule, i.e., we explored maps of different sizes to avoid

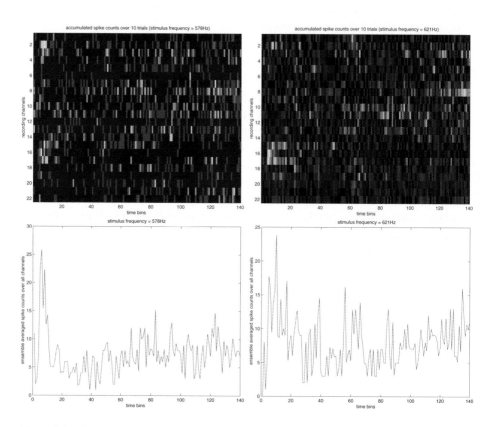

Figure 5.2. Top row: accumulated spike counts over 10 trials for 22 selected neurons in the guinea pig auditory cortex. The horizontal axis represents one complete trial: it takes 200ms from stimulus onset to offset and the interstimulus interval is 500ms. Time bins of 5ms are used in computing the spike rates. Bottom row: spike counts accumulated over all channels from the top row. The left column corresponds to a stimulus frequency of 578 Hz and right of 621Hz.

over-fitting or less-fitting. We have chosen the training parameters in the SOM algorithm to be $\alpha(k) = 0.02$ in beginning and gradually falls down to 0.0001 following a Gaussian curve during the training phase. This parameter then is reduced to minimum in the fine tuning phase. The neighborhood function $N_c(k)$ includes all the output nodes in the beginning and it reduces to 0 at the end of training. The reduction in the size of the neighborhood also follows a Gaussian curve. Half way through the training process the neighborhood size is retained to about 40% of the original size.

To train the network, we select n cells (in this study, n=11) in the auditory cortex, representing the most active cells among recorded. The average discharge rates of these n cells constitute a discharge rate vector $X_j = [d_{1j}, d_{2j}, \cdots, d_{nj}]^T$, where $d_{ij}, i = 1, \cdots, n$, is the average discharge rate of the ith cell at time bin t_j. The X_j's are then lumped together from the first to the last bin to form a bigger input vector for the SOM training. Throughout this study, data from 10 complete trials were used for analysis. We therefore have used *Leave-One-Out* to obtain our training and testing results. Or specifically, among the 10 trials, 9 of which were used for training the SOM and the 1 trial left was used for testing.

Before the SOM training, further data preparation was performed. Specifically, we examined the individual channel waveforms and realized that for 9 out of the 60 frequency stimuli, there was little cortical response in almost all of the 33 electrodes. These 9 frequency stimuli were then excluded from further analysis since there is no reliable carrier signal. We then constructed 5 SOMs, each was used for classifying a sub-group of the 45 frequency stimuli. This is to increase the training and testing accuracy and also to reduce CPU time. After this preparation, for each of the SOM network, we have an input vector of size $11 \times$ (the number of bins used). And for each SOM, we have 81 training vectors and 9 testing vectors for each *Leave-One-Out* exercise.

During training, a vector $X \in \{X_k | k = 1, 2, \cdots, 81\}$ is selected from the data set at random. The discharge rate vectors are not labeled or segmented in any way in the training phase: all the features present in the original discharge rate patterns will contribute to the self-organization of the map. Once the training is done, as described previously, the network has learned the classification and topology relations in the input data space. Such a "learned" network is then calibrated using the discharge rate vectors of which classifications are known. For instance, if a node i in the output layer wins most for discharge rate patterns from a certain frequency stimulus, then the node will be labeled with this frequency.

Table 5.1 summarizes the analysis performed by one of the five SOMs on 10 auditory trials for each of the stimulus frequency. In the following results, 50ms bins are used and the first 8 bins of each trial are used as inputs to the SOM. Similar results were also obtained by smaller bin sizes such as 40ms or 20ms.

Further care will be given if smaller bin sizes are to be used in the current SOM framework since the number of trials, or consequently the data set, is relatively small, compared to the SOM network size.

Table 5.1. Typical frequency discrimination results using one of the five SOM networks

SOM size	Epochs [a]	Train/Test accuracy [b]	LOO accuracy [c]
10×10	180+250	77.8/44.4	30.0
15×15	200+1000	77.8/55.6	32.2

[a] Traning + Fine tuning epochs.
[b] Traning accuracy (%) and representative good testing accuracy.
[c] Averaged testing accuracy using *Leave-One-Out*.

Compared with the successful classification rate using simple template matching, the SOM has significantly improved the classification accuracy. These classification rates are much higher than the probabilities if classes are chosen by chance (which is 1.67%). Recall that the data we used were obtained from a non-behaving animal. Our results reveal that there is strong correlation between external stimuli and the neuronal firing patterns in the neuron ensemble.

5.   Conclusions and Discussions

This chapter has demonstrated the effectiveness of the SOM algorithm in complex data clustering and data classification tasks. As simultaneous cortical recording becomes more readily accessible to neuroscientists, the need for developing and validating efficient spike train decoding algorithms is apparent. In this relatively new but challenging area, several statistical tools have been applied to spike train analysis. Some are motivated by neurophysiology, such as population vector analysis [45], whereas others use more "standard" statistical tools, such as maximum likelihood, maximum entropy, principal component analysis, etc. One major difficulty with many of these algorithms is the need to estimate the probability density function (which requires a large number of trials) before any hypothesis testing can be performed. Some of the algorithms have proven to be optimal only when the input data distribution is Gaussian or when the data set is linearly separable. These conditions are not often valid using data from multichannel cortical recording.

Our analysis of the guinea pig data reveals that SOM has provided a much higher level of discrimination among the recorded signals than discrimination by chance or by template matching. The next step is to apply the SOM to decode neural firing patterns in response to behaviorally relevant stimuli (e.g., animals doing frequency discrimination) and more complex sets of acoustic stimuli.

Acknowledgments

The research is supported by DARPA under grant MDA 972-00-1-0027 and by NSF under grant ECS-0002098.

References

[1] G. G. Blasdel and G. Salama, "Voltage-sensitive dyes reveal a modular organization in monkey striate cortex," *Nature*, vol. 321, pp. 579-585, 1986.

[2] D. H. Hubel and T. N. Wiesel, "Receptive fields, binocular interaction and functional architecture in the cat's visual cortex," *Journal of Physiology*, (London), vol. 160, pp. 106-154, 1962.

[3] D. H. Hubel and T. N. Wiesel, "Sequence regularity and geometry of orientation columns in the monkey striate cortex," *Journal of Comparative Neurology*, vol. 158, pp. 267-294, 1974.

[4] E. I. Knudsen, S. du Lac, and S. D. Esterly, "Computational maps in the brain," *Annual Review of Neuroscience*, vol. 10, pp. 41-65, 1987.

[5] G. M. Shepherd, *Neurobiology*, 2nd edition. New York: Oxford University Press, 1988.

[6] C. von der Malsburg, "Self-organization of orientation sensitive cells in the striate cortex," *Kebernetil*, vol. 14, pp. 85-100, 1973.

[7] D. J. Willshaw and C. von der Masburg, "How patterned neural connections can be set up by self-organization," *Proceeding of the Royal Society of London*, Series B 194, pp. 431-445, 1976.

[8] T. Kohonen, "self-organized formation of topologically correct feature maps," *Biological Cybernetics*, vol. 43, pp. 59-69, 1982,

[9] T. Martinetz, H. Ritter, and K. Schulten, "Three-dimensional neural net for learning visuomotor coordination of a robot arm," *IEEE Trans. on Neural Networks*, vol. 1, no.1, pp. 131-136, 1990.

[10] H. Ritter and K. Schulten, "Topology conserving mappings for learning motor tasks," *AIP Conference Proceedings*, vol. 151, pp. 376-380, Snowbird, Utah, 1986.

[11] H. Ritter and K. Schulten, "Topology-conserving maps for learning visuo-motor-coordination," *Neural Networks*, vol. 2, pp. 159-168, 1989.

[12] M. Kasslin, J. Kangas, and O. Simula, "Process state monitoring using self-organizing maps," In I. Aleksander, J. Taylor (Editors), *Artificial Neural Networks, 2*, pp. II-1531-1534, North-Holland, September 1992.

[13] V. Tryba and K. Coser, "Self-organizing feature maps for process control in chemistry," In T. Kohonen, K. Makisara, O. Simula, and J. Kangas (Editors), *Artificial Neural Networks*, pp. I-874-852, North-Holland, 1991.

[14] H. Ultsch, "Self-organizing feature maps for monitoring and knowledge acquisition of a chemical process," In S. Gielen and B. Kappen (editors), *Proceedings of 1993 International Conference on Artificial Neural Networks (ICANN-93)*, pp. 864-867. London: Springer-Verlag, 1993.

[15] M. Abeles, *Corticonics*, Cambridge University Press, 1991.

[16] T. Kohonen, "The self-organizing map," *Proceedings of the IEEE*, vol.78, no. 9, pp. 1464-1480, 1990.

[17] R. Togneri, M.D. Alder, and Y. Attrikiouzel, "Dimension and structure of the speech space," *IEE proceedings-I, Commun. Speech & Vision*, vol. 139, no. 2, pp. 123-127, 1992.

[18] B. Zhu, M. Ramsey, and H. Chen"Creating a large-scale content-based airphoto image digital library," *IEEE Transactions on Image Processing*, Volume: 9 Issue: 1 , pp. 163 -167, Jan. 2000.

[19] T. Kohonen, S. Kaski, K. Lagus, J. Salojarvi, J. Honkela, V. Paatero, and A. Saarela, "Self organization of a massive document collection," *IEEE Transactions on Neural Networks*, Volume: 11 Issue: 3 , pp. 574 -585, May 2000.

[20] O. Yanez-Suarez, and M. R. Azimi-Sadjadi, "Unsupervised clustering in Hough space for identification of partially occluded objects," *IEEE Transactions on Pattern Analysis and Machine Intelligence*, Volume: 21 Issue: 9 , pp. 946 -950, Sept. 1999.

[21] B. Zhang, M. Fu; H. Yan, and M. A. Jabri, "Handwritten digit recognition by adaptive-subspace self-organizing map (ASSOM)," *IEEE Transactions on Neural Networks*, Volume: 10 Issue: 4 , pp. 939 -945, July 1999.

[22] F-L Chen, and S-F Liu, "A neural-network approach to recognize defect spatial pattern in semiconductor fabrication," *IEEE Transactions on Semiconductor Manufacturing*, Volume: 13 Issue: 3 , pp. 366 -373, Aug. 2000.

[23] D. Niebur and A. J. Germond, "Power system static security assessment using the Kohonen neural network classifier," *IEEE Trans. on Power System*, vol. 7, no. 2, pp. 865-872, 1992.

[24] G. Wittington and T. Spracklen, "The application of a neural network model to sensor data fusion," *Proceeding of SPIE –Applications of Artificial Neural Networks*, vol. 1294, pp. 276-283, 1990.

[25] W. Wan, and D. Fraser, "Multisource data fusion with multiple self-organizing maps," *IEEE Transactions on Geoscience and Remote Sensing*, Volume: 37 Issue: 3 Part: 1 , pp. 1344 -1349, May 1999.

[26] S. Lin, J. Si, A. B. Schwartz, and G. T. Yamaguchi, "Self-Organization modeling of firing activities in primate motor cortex," *Society for Neuroscience Abstracts*, vol. 21, pp. 515, Nov. 1995.

[27] S. Lin, J. Si, and A. B. Schwartz, "Self-organization of firing activities in monkey's motor cortex: trajectory computation from spike signals," *Neural Computation,* vol. 9, no. 3, pp. 607-621, 1997.

[28] J. Si, D. R. Kipke, S. Lin, A. B. Schwartz, and P. D. Perepelkin, "Motor Cortical Information Processing," in H. Eichenbaum and J. L. Davis (eds.), *Strategies in the Study of Biological Neural Networks*. John Wiley and Sons, Inc. New York, pp. 137-159, 1998.

[29] A. J. Hoglund, K. Hatonen, and A. S. Sorvari, "A computer host-based user anomaly detection system using the self-organizing map," *Proceedings of the IEEE-INNS-ENNS International Joint Conference on Neural Networks*, Volume: 5 , pp. 411 -416, 2000.

[30] S. B. Garavaglia, "Health care customer satisfaction survey analysis using self-organizing maps and 'exponentially smeared' data vectors," *Proceedings of the IEEE-INNS-ENNS International Joint Conference on Neural Networks*, Volume: 4 , pp. 119 -124, 2000.

[31] H. Douzono, S. Hara, and Y. Noguchi, "A clustering method of chromosome fluorescence profiles using modified self organizing map controlled by simulated annealing," *Proceedings of the IEEE-INNS-ENNS International Joint Conference on Neural Networks*, Volume: 4 , pp. 103 -106, 2000.

[32] E. Oja, "A simplified neuron model as a principle component analyzer," *Journal of Mathematical Biology,* vol. 15, pp. 267-273, 1982.

[33] T. Kohonen, *Self-organizing map*. Heidelberg: Springer-Verlag, 1995.

[34] T. Kohonen, *Self-organization and associative memory*. Springer Series In Information Sciences, Berlin : Springer-Verlag, 2nd edition, 1988.

[35] H. Robbins and S. Monro, "A stochastic approximation method," *Ann. Math. Stat.*, vol. 22, pp. 400-407, 1951.

[36] S. Lin and J. Si , "Weight value convergence of the SOM algorithm," *Neural Computation*, vol. 10, 807-814, 1998.

[37] Y. Linde, A. Buzo, and R. M. Gray, "An algorithm for vector quantizer design," *IEEE Trans. on Communications,* vol. COM-28, no. 1, pp. 84-95, 1980.

[38] J. W. Sammon, "A nonlinear mapping for data structure analysis," *IEEE Trans. Comput.,* vol. C-18, pp. 401-409, 1969.

[39] J. Mao and A. K. Jain, "Artificial neural networks for feature extraction and multivariate data projection," *IEEE Trans. on neural networks*, vol. 6. no. 2, pp. 296-317, 1995.

[40] S. Haykin, *Neural Networks: a comprehensive foundation*, 2nd Edition, New York: MacMillian College Publishing Company, 1999.

[41] E. Oja, "Neural networks, principal components, and subspaces," *Journal of Neural Systems,* vol. 1, pp. 61-68, 1989.

[42] C. Jutten and J. Herault, "Blind separation of sources, part I: an adaptive algorithm based on neuromimetic architecture," *Signal Processing,* vol. 24, no. 1, pp. 1-10, 1991.

[43] M. L. Sutter, and C. F. Schreiner, "Physiology and topography of neurons with multipeaked tuning curves in cat primary auditory cortex," *Journal of Neurophysiology.* 65: 1207-1225, 1991.

[44] F. Rieke, D. Warland, R. de Ruyter van Steveninck, and W. Bialek, *Spikes: Exploring the Neural Code*, the MIT Press, 1996.

[45] A. B. Schwartz, " Direct cortical representation of drawing," *Science*, Vol. 265, pp. 540-542, 1994.

Chapter 6

FINDING TRANSITION STATES USING THE LTP ALGORITHM

Cristian Cardenas-Lailhacar
Industrial and Systems Engineering,
University of Florida
303 Weil Hall, P.O. Box 116595,
Gainesville, Florida 32611-6595

Michael C. Zerner
Quantum Theory Project,
University of Florida,
Gainesville, Florida 32611-8435

Abstract   Preliminary results using the Line-Then-Plane (LTP) procedure to find Transition States (TS) in simple chemical systems are here presented. With slight modifications the algorithm has been implemented in the ZINDO program package, and has proven to require less computer time than other procedures that demand the full evaluation of the Hessian. We demonstrate that this procedure is able to give a good approximation of the minimum energy reaction path.

1.   Introduction

The application of transition state search techniques has a significant impact in a myriad of research disciplines concerning the following: Nanotechnology, biodiagnostics, self-assembly in devices or nanotechnological elementary machines, and biomolecular computers (electronic devices which can be combined in biocomputing architectures). Other major research and application areas include drug design, industrial chemical processes, catalysis, solid state physics, most dynamical solid materials property studies, environmental studies (as the reactions involved in the ozone layer depletion), etc.

P.M. Pardalos and J. Principe (eds.), Biocomputing, 107-128.

One of the great advances in quantum chemistry in the last 25 years is the introduction of gradient methods in searching potential energy surfaces (PES) for the stable structure of molecules given only the atomic nuclei and the number of electrons. Although formidable problems still remain with larger molecules with many minima and many soft vibrational modes, finding minima is now relatively straightforward. Given the Hessian, or second derivative matrix, these stable structures can easily be characterized by their calculated vibrational spectra.

A much harder problem is characterizing a molecule relative to re-arrangements to other isomeric forms, or to reactions with other molecules. This generally requires the determination of a transition state (TS), an extreme point on the surface with one and only one negative eigenvalue of the Hessian matrix, whose eigenvector corresponds to the reaction coordinate. Transition states are well defined quantum chemically even if they have only ephemeral existence experimentally. Rather effective methods exist for determining transition states, provided that steps from the reactant (**R**) toward the product (**P**) are modest enough, and they are guided with knowledge of both gradient and Hessian. However, the calculation of the Hessian at each point along the path is computationally expensive, limiting such a procedure to small molecular systems only. Perhaps the most successful and feasible general method for determining a TS is to guess, and then search using Newton methods to find the nearest extreme point. Unfortunately our intuition in guessing a transition state is far less developed than is our ability to guess minima (stable structures) and we often uncover extreme points in the surface that are other minima, or that have two or more negative Hessian eigenvalues, indicating yet a lower energy path between reactant and product. It appears that an evaluation of the Hessian, either using analytic or numerical techniques, will always be required to ensure that a point is a TS, but we would like to avoid the constant evaluation of the Hessian as we proceed from reactants to products.

However, the most efficient algorithms to find TS's use second derivative matrices [1-16], requiring a great computational effort, but none of these is as yet generally successful. Unfortunately, and after more than two decades of study, effective algorithms to solve this problem inexpensively using computational methods are still not available. This fact alone is a powerful incentive to try to develop new procedures that do not require the Hessian.

With the exception of the Synchronous Transit [1] and the Normalization Technique [16] models, generally all other procedures uses up-hill methods to search for the transition state starting from reactants through a second order expansion of the energy in terms of a Taylor series.

In a previous work we have explored a Line-Then-Plane (LTP) algorithm [17] that does not require the evaluation of the Hessian as we search in the direction from reactants to products and simultaneously from products to reactants. This was checked against six of the most commonly used analytical functions used as a reference for these kinds of searching algorithms. In each case, the search was successful and competitive with procedures that require the analytical evaluation of the Hessian. We here check this procedure on real chemical systems using the model Intermediate Neglect of Differential Overlap (INDO) Hamiltonian in the ZINDO program package [18] at the SCF level, although such a procedure is, of course, limited to no particular energy function, provided it is continuous.

The LTP procedure determines a series of points representing products, $\mathbf{P}_i$, and reactants $\mathbf{R}_i$ along lines connecting $\mathbf{P}_i$ and $\mathbf{R}_i$, as discussed in more detail later. A step is taken and then a minimization in hyperplanes perpendicular to this direction using quasi-Newton update techniques such as the BFGS procedure [19]. If the LTP procedure is to be successful, then the many searches in these perpendicular planes that do not use Hessians, each representing full energy calculations, must be competitive in computational time with the evaluation of the Hessian at each point $\mathbf{P}_i$ and $\mathbf{R}_i$. Such searches are certainly advantageous over the evaluation of second derivatives in terms of computer storage requirements.

Results of the LTP procedure to find TS in five simple chemical reactions are presented here. The results are discussed and compared with the ones obtained through the Augmented Hessian method [20], also implemented in the ZINDO program package, in terms of accuracy and computer memory storage. The main focus of this work is to test the algorithm as well as to prove the ease of evaluating the energy and gradient in $N - 1$ directions rather than evaluate $N(N + 1)/2$ second derivatives, where N is the number of coordinates to be searched. In the next section a brief theoretical background is given. In the third section the fundamental relations of the LTP procedure are presented. We summarize and discuss our results in the fourth part.

2. Theoretical background

In searching for extremal points in a Potential Energy Surface (PES) the energy (E) and gradient (f) are usually expanded as a Taylor Series around a given geometry x, truncated at second order,

$$E(x) = E(x_o) + qf^\dagger + \frac{1}{2}q\mathbf{H}q^\dagger \tag{1}$$

and,

$$f = f_o + q\mathbf{H} \tag{2}$$

From here,

$$q = -f\mathbf{H}^{-1} \tag{3}$$

Also,

$$s = \alpha q \quad \text{where:} \quad q = X - X_o \qquad (4)$$

Where s is the step which is a fraction of the variation of coordinates q. Minima, Transition States (TS) and Intermediates of Reaction, all extremal points, are given by q when $f_o = 0$.

The Line Search Technique to find the best value of α is described elsewhere [22 - 25]. Updating procedures for the Hessian $\mathbf{H}$ or its inverse $\mathbf{G} = \mathbf{H}^{-1}$ has been widely investigated. The transition state is said to be found [27] when: **1)** q is a stationary point, **2)** the Hessian has one and only one negative eigenvalue, **3)** it has the highest energy in the line connecting reactants and products and **4)** it represents the lowest energy in all other directions.

3.   The LTP procedure

We report here the first results using the *Line-Then-Plane* approach for finding Transition States in Chemical Reactions. We do not describe the **LTP** procedure in great detail here, but rather refer to our previous work [17]. For completeness, however, we do give some details so that the relations governing this algorithm are reasonably complete.

The ***Exact LTP*** procedure [17], in which the BFGS technique [19] is used to minimize the energy in the hyperplanes perpendicular to the directions $\mathbf{d}_i$, connecting reactants $\mathbf{R}_i$ and products $\mathbf{P}_i$, has been implemented in the ZINDO program package [18] and tested in five simple chemical reactions.

The structures of reactants Ri and products Pi are optimized. The displacement $\mathbf{d}_i = \mathbf{P}_i - \mathbf{R}_i$ is defined and the step is a fraction of the way from $\mathbf{R}_i$ to $\mathbf{P}_i$ along $\mathbf{d}_i$, and from $\mathbf{P}_i$ to $\mathbf{R}_i$ along $-\mathbf{d}_i$. The structures of these two points will be **minimized** in the hyperplanes (*i.e.*all directions) perpendicular to $\mathbf{d}_i$, defining the new points $\mathbf{R}_{i+1}$ and $\mathbf{P}_{i+1}$. Then a new difference vector $\mathbf{d}_{i+1} = \mathbf{P}_{i+1} - \mathbf{R}_{i+1}$ is defined and the procedure repeated until convergency criteria are realized.

3.1.   The strategy

We adopt the following strategy:

Step 1) Optimize the Reactant and Product geometries $\mathbf{R}_o^p$ and $\mathbf{P}_o^p$ (the superindex p stands for projected coordinates) Set counters $i = j = 0$ (see below).

Step 2) Define the displacement vector $\mathbf{d}_i = \mathbf{R}_i^p - \mathbf{P}_i^p$. If the norm $\mathbf{d}_i^\dagger \mathbf{d}_i < T$ (a given threshold), stop. Otherwise,

Step 3) Examine $\sigma^\dagger(\mathbf{R}_{i-1}^p)\mathbf{d}_i$ and $-\sigma^\dagger(\mathbf{P}_{i-1}^p)\mathbf{d}_i$, where $\sigma(x) = (\partial E/\partial \mathbf{d}_i)_x$. Changes in the sign of this parameter are registered. The projected coordinates are set to the line (l) connecting $\mathbf{R}_i$ and $\mathbf{P}_i$

$$\text{Set} \quad \mathbf{R}_i^l = \mathbf{R}_i^p; \quad \mathbf{P}_i^l = \mathbf{P}_i^p \quad i = 1, 2, \ldots \tag{5}$$

(the superindex l stands for variables *in the line* that connects the corresponding R's and P's) unless,

$$i) \qquad \text{If} \; \sigma^\dagger(\mathbf{R}_i^p)\mathbf{d}_i < 0 \; \text{ set } \; \mathbf{P}_i^l = \mathbf{R}_i^p \; \text{ and } \; \mathbf{R}_i^l = \mathbf{R}_{i-1}^l \tag{6}$$

or,

$$ii) \qquad \text{If} \; -\sigma^\dagger(\mathbf{P}_i^p)\mathbf{d}_i < 0 \; \text{ set } \; \mathbf{R}_i^l = \mathbf{P}_i^p \; \text{ and } \; \mathbf{P}_i^l = \mathbf{P}_{i-1}^l \tag{7}$$

Step 4) Walk along $\mathbf{d}_i$ from $\mathbf{R}_i^l$ to $\mathbf{P}_i^l$, and *vice-versa*, a fixed step of length $s_i = d_i/N_i$

$$\text{Set, If} \; \mathbf{R}_{i+1}^l = \mathbf{R}_i^l + s_i \; \text{ and } \; \mathbf{P}_{i+1}^l = \mathbf{P}_i^l - s_i \tag{8}$$

Unless,

$$\text{If} \; T < \mathbf{d}_i^\dagger \mathbf{d}_i \; \leq s_j^\dagger s_j \; \text{set} \; N_i = N_i/2 \; \text{ and } \; j = i \tag{9}$$

$$\text{If} \; N_i < 2 \; \text{set} \; N_i = 2 \tag{10}$$

Step 5) As in Step 3), examine $\sigma^\dagger(\mathbf{R}_{i+1}^l)s_i$ and $-\sigma^\dagger(\mathbf{P}_{i+1}^l)s_i$:

$$\text{If} \; \sigma^\dagger(\mathbf{R}_{i+1}^l)s_i < 0 \; \text{ set } \; \mathbf{P}_{i+1}^l = \mathbf{R}_{i+1}^l \; \text{ and } \; \mathbf{R}_{i+1}^l = \mathbf{R}_i^l \tag{11}$$

$$\text{If} \; -\sigma^\dagger(\mathbf{P}_{i+1}^l)s_i < 0 \; \text{ set } \; \mathbf{R}_{i+1}^l = \mathbf{P}_{i+1}^l \; \text{ and } \; \mathbf{P}_{i+1}^l = \mathbf{P}_i^l \tag{12}$$

Step 6) Minimize in the hyperplanes perpendicular to $\mathbf{d}_i$ containing $\mathbf{R}_{i+1}^l$ and $\mathbf{P}_{i+1}^l$ to obtain new projected geometries $\mathbf{R}_{i+1}^p$ and $\mathbf{P}_{i+1}^p$, respectively. Set $i = i + 1$ and go to Step 2).

Although it can be shown that the LTP technique must lead to the Transition State on a continuous potential energy surface E(x) if the steps are conservative enough, the norm of the gradient (f) and the Hessian $(\mathbf{H})$ are examined at convergence in order to ensure that the converged point on E(x) has the right inertia. (*i.e.*, $f = 0$, and $\mathbf{H}$ with one and only one negative eigenvalue).

To accelerate convergence to the TS, we might add to Step 3) a further test that becomes useful as the TS is approached,

Step 3) iii)

$$If\ \Delta\sigma^{\dagger}\mathbf{s}(k) = \sigma^{\dagger}(\mathbf{P}_{k+1}^l)\mathbf{s}_k - \sigma^{\dagger}(\mathbf{R}_{k+1}^l)\mathbf{s}_k < T'\quad and,$$
$$\Delta E(k) = E(\mathbf{P}_k) - E(\mathbf{R}_k) < T'', k = i, i - 1 \qquad and\ i - 2 \tag{13}$$

Then, set: $q_i = (\mathbf{P}_i - \mathbf{R}_i)/2$, evaluate $E(q_i)$, $f\ (q_i)$ and $\mathbf{H}(q_i)$
Else, If

$$f^{\dagger}(q_i)\quad f(q_i) < f^{\dagger}(\mathbf{P}_i)\quad f(\mathbf{P}_i)\quad and\quad -f^{\dagger}(q_i)\mathbf{d}_i < 0 \tag{14}$$

Then q_i replaces $\mathbf{R}_i$
Then, go to Step 2), *Else*, go to Step 4).

Here q_i refers to coordinates that represent a conformation that is very close to the TS structure. Based on our experience, the choice N = 10 generates a conservative initial step and suitable thresholds are $T = T' = T'' = T''' = 10^{-4}$ arbitrary units.

The strategy delineated above is also successful even for systems which have intermediate structures between R and P. The tests indicated in equations (6) and (7), under Step 3), and (11) and (12), under Step 5), disclose potential turning points caused either by too large a step from **R** toward **P** or **P** toward **R**.

The reaction path can be approximated by connecting all points $\mathbf{R}_i$ and $\mathbf{P}_i$. An approximate and faster procedure would be to quit in Steps 3) or 5), thereby avoiding the rest of coordinates between consecutive steps. Then the displacement $\mathbf{d}_i$ can now be divided in smaller parts (say 4) and the procedure continued as before. The last half is now submitted to a perpendicular minima line search founding a last point X_e, the TS.

In general, the TS is said to be found if the gradient norm is neglegible and if the Hessian has one and only one negative eigenvalue, respectively. As for LTP, the transition state will be considered to be found when the norm of the displacement vector $\mathbf{d}_i$ is smaller than a pre-established threshold T_c (usually $T_c < 10^{-3}$). Nevertheless, the general conditions are checked at the estimated saddle point (i.e. $\sigma(X_e,) = 0$).

3.2.   The projected coordinates

The coordinates perpendicular to the direction $\mathbf{d}_i$ are obtained by projection, and the energy in the hyperplane minimized using the BFGS algorithm as developed by Head and Zerner [23]. This is

$$\mathbf{P}_{di}(x_{i+1} - x_{i+1}^l) = \mathbf{P}_{di}(x_{i+1} - s_{i+1}) = -\alpha\mathbf{P}_{di}\mathbf{G}_{i+1}\mathbf{P}_{di}f_{i+1} \tag{15}$$

Where $s_{i+1} = d_i / N = x_{i+1}^l$

is the step (coordinates) along the line connecting projected R's and P's.

$$x_{i+1}^{pi} = x_{i+1}^{l} - \alpha \mathbf{G}_{i+1}^{pi} f_{i+1}^{pi} \tag{16}$$

Where α is the line search parameter that determines how far along the direction of the step one should proceed, (in the quadratic region of the PES α = 1). We use the value α_i = 0.3 for all i, which is a more conservative value than that recommended by Head, Weiner and Zerner [21, 24] (α_o = 0.4 and all other α_i = 1.0).

3.3. The projector

The projector perpendicular to d_i is defined as [17, 21]:

$$\mathbf{P}_{di} = \mathbf{I} - \mathbf{d}_i \mathbf{d}_i^{\dagger} / \mathbf{d}_i^{\dagger} \mathbf{d}_i \tag{17}$$

Coordinates to obtain the displacement vector should come in internal coordinates rather than in cartesians so as to avoid translations and rotations in each perpendicular search plane.

LTP requires knowledge of reactants ($\mathbf{R}_0$) and products ($\mathbf{P}_0$); no previous knowledge of the TS is necessary. It makes partial use of the line search technique and the search for minima in perpendicular directions is performed using the BFGS technique. The initial search direction is from reactant to product. Since the initial step in this direction is modest, minimization in the perpendicular direction allows considerable deviation from the initial direction.

The main difference of LTP with most available procedures, is that the more stable structures of products and reactants, the direction in which to walk, a constrained step size **s** and a simultaneous walking from **R** and **P** towards the TS along a line connecting them, and a free of translations and rotations projector are considered. The concept of considering **R** and **P** approaching the TS simultaneously is mainly based on the idea that one contains information, history, about the other having in common the same TS, as if we were thinking of the energy barrier to be the lowest passage from one side of a mountain to the other (**R** to **P** and *vice-versa*). The advantage of this idea is that we look to the PES (mountain) from both sides, and not from one as usual. Finally, and because of the Newton-Raphson nature of the LTP steps [26], the procedure will unequivocally converge to the TS on any continuous surface.

4. Results and discussion

The LTP procedure was previously tested successfully on six model potential functions [17] often used for the purpose of examining TS searching procedures. Recently implemented in the ZINDO program package, we tested our procedure on five molecular systems in a first attempt to demonstrate the utility of the LTP method in finding transition states in 3N - 6 dimensional systems. The lower computational time required, as well as the small number of SCF calculations needed to find the TS, confirms the basis of our procedure. Namely that the single evaluation of the Hessian at the found TS point shows the procedure's benefit in terms of ease, lower computational time, as well as, in its ability to approximate the lowest energy reaction path.

Our results suggests that, for the treatment of the Hessian, update procedures must be used provided that they do not change its natural signature. On the suggested Transition State, which in LTP is given by the square of the norm of the gradient and its maximum component smaller than a given threshold along with the slope approaching zero, an accurate final geometry can be obtained by evaluating the exact Hessian.

Ammonia Isomerization Reaction

Because of the size of this molecular system, the isomerization reaction of ammonia has been one of the most used reactions to study the behavior of TS search algorithms. To better test and check LTP, we have considered three initial configurations for ammonia isomerization reactions. First we consider the symmetric case in which reactants and products are nothing but a mirror image of each other. Next we study an asymmetric configuration where now the distance from the nitrogen atom to the hyperplane formed by the hydrogen atoms is smaller than the one of the reactants. In the third case we study the effect of rotations and translations by a 50° rotation of the product ammonia molecule of the first case, with respect to the reactants. The three situations including initial geometry parameters are shown in Figure 6.1. Detailed results for the symmetric case are displayed for both reactants and products in Table 6.1. In Table 6.2 we display our results for all three cases including Exact-LTP, Approximate LTP, and a geometry optimization on the approximate procedure. Also shown are those obtained using the Augmeneted Hessian technique (also implemented in the ZINDO program package).

Symmetric Inversion of Ammonia

Figure 6.1-a shows a 3-dimensional scheme for the inversion of ammonia, including both the initial reactants and products initial geometries as well as the TS found. A detailed LTP cycle-by-cycle uphill walk which includes energy, geometry and the number of iterations used to update the Hessian, are summarized in Table 6.1.

Our results are collected and compared with those coming from the Augmented Hessian (AH) technique, in Table 6.2. They show that the TS found LTP is, energetically and geometrically, the same as the one found using the AH method. The main difference of both procedures is that AH uses analytical

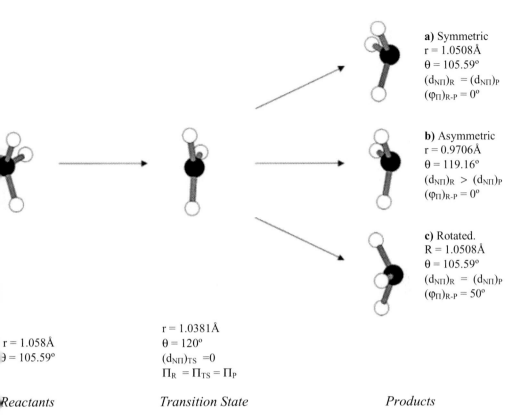

a) Symmetric
$r = 1.0508\text{Å}$
$\theta = 105.59°$
$(d_{N\Pi})_R = (d_{N\Pi})_P$
$(\varphi_\Pi)_{R-P} = 0°$

b) Asymmetric
$r = 0.9706\text{Å}$
$\theta = 119.16°$
$(d_{N\Pi})_R > (d_{N\Pi})_P$
$(\varphi_\Pi)_{R-P} = 0°$

c) Rotated.
$R = 1.0508\text{Å}$
$\theta = 105.59°$
$(d_{N\Pi})_R = (d_{N\Pi})_P$
$(\varphi_\Pi)_{R-P} = 50°$

$r = 1.0381\text{Å}$
$\theta = 120°$
$(d_{N\Pi})_{TS} = 0$
$\Pi_R = \Pi_{TS} = \Pi_P$

$r = 1.058\text{Å}$
$\theta = 105.59°$

Reactants   *Transition State*   *Products*

Figure 6.1.   The three variations of the Inversion reaction of Ammonia studied here are shown: **a)** Symmetric, **b)** Asymmetric, and **c)** Rotated. The nitrogen-hydrogen bond length (r), the hydrogen-nitrogen-hydrogen angle θ, and the distance (d) from the nitrogen to the hydrogen atoms plane (Π) are shown. φ is the angle by which the hydrogen plane of the products have been rotated with respect to that of the reactants (**c**).

second derivatives and consequently it better represents the uphill path on the PES using 7 SCF iterations. On the other hand, LTP updates the Hessian using the BFGS [20] technique, and walks uphill towards the TS from both *sides* of the reaction, this is, reactants and products, but with 22 cycles.

Notice that the Approximate LTP gives a reasonable good answer without losing much in accuracy, using a number of SCF cycles (10) markedly less than those used by the Exact LTP [17]. However, the App-LTP has the intrinsic problem of not containing enough information about the topography of the PES. Consequently, it requires a final geometry optimization (App-LTP-GOPT) at the TS. When this last calculation is performed, only 3 extra SCF cycles are required. A substantial improvement is achieved in this way, as is shown in Table 6.2. Performing this final optimization is not risky at all because the TS

Table 6.1. LTP geometry, energy, and number of iterations required to find the TS for the symmetric inversion of NH_3. The results are from the Exact Line-Then-Plane technique. The energy (in Hartrees), the optimized geometry in the plane perpendicular to the walk direction (namely the nitrogen-hydrogen bond length (r in Angstroms)), and the hydrogen-nitrogen-hydrogen angle (θ, in degrees), the distance (d, in Angstroms) between R's and P's, the maximum component of the gradient (f_{max}) and the number of update iterations (UI) of the Hessian at each LTP cycle are displayed. The convergence criteria were that the maximum component of the gradient and the distance to be smaller than 10^{-3} (Hartrees/Angstroms). The total number of energy evaluations was 31. One of the acceleration flags of LTP is turned on. The step factor is now amplified by 4. At this LTP cycle, and as the distance d (the square of the norm of the displacement vector) is smaller than 0.1, a threshold pre-established by the LTP algorithm, the step is amplified by a factor of 2.

	Energy	r_{N-H}	θ_{HNH}	d_I	f_{max}	UI
$R_o = P_o$	-12.522855	1.0503	105.63°	1.1825	0.000000	0
$R_1 = P_1$	-12.520988	1.0452	110.54°	0.9460	0.000059	3
$R_2 = P_2$	-12.517629	1.0421	113.83°	0.7568 [a]	0.000627	2
$R_3 = P_3$	-12.507616	1.0383	119.74°	0.1514	0.000206	3
$R_4 = P_4$	-12.507098	1.0381	119.99°	0.0303 [b]	0.000480	2
R_5 step	-12.507077	1.0381	120.00°		0.000480	0

found by the App-LTP has the right signature over the Hessian as it lies in the surroundings of the TS.

Asymmetric Inversion of Ammonia

Figure 6.1-b shows the geometry of the initial reactants and products as well as for the TS found. The results are displayed in Table 6.2. It is noticeable that the TS found by LTP is as good as the one coming from the AH technique. The differences between both procedures have been already described. Notice that the same trend of results obtained in the previous case is here obtained too, with the caveat that the App-LTP now gives a much better estimate of the TS, whose final optimization now requires only 2 additional SCF cycles.

Rotated Symmetric Inversion of Ammonia

This test was performed to establish whether the LTP procedure is capable of reaching the well-known planar TS conformation of ammonia, by means of a rotation through 50 degrees of the inverted umbrella structure (products). The rotation is relative to the umbrella conformation (reactants), in the plane that contains the hydrogen atoms as shown in Figure 6.1-c, through a step-by-step procedure during the up-hill walk. The results are shown in Table 6.2.

The results displayed in Table 6.2 show that with only a few more SCF calculations than before, LTP (exact) is not only able to rotate the reactants

Table 6.2. LTP geometry, energy, and number of iterations required to find the TS to the Symmetric, Asymmetric, and Rotated Symmetric NH_3 Isomerization reactions. For each of these reactions, we show results obtained using the Augmented Hessian (AH) technique [20], the Exact and Approximate LTP techniques. A final geometry optimization calculation performed on the TS found by the Approximate LTP procedure (App-LTP-GOPT) is also shown. The nitrogen-hydrogen bond length r is in Angstroms, θ represents the hydrogen-nitrogen-hydrogen angle (in degrees), and E is the Energy (in Hartrees). Finally the number of SCF cycles used by each procedure are also displayed. The convergence criteria were for both, the norm of the gradient and the displacement, to be smaller than 10^{-3}.

Inversion of NH_3	AH	Exact-LTP	App-LTP	App-LTP-GOPT
Symmetric				
r	1.0382	1.0382	1.004	1.0380
θ	120°	120°	120°	120°
E	-12.507084	-12.507079	-12.498321	-12.507076
Cycles	7	22	10	3
Asymmetric				
r	1.0382	1.0380	1.0367	1.0381
θ	120°	120°	120°	120°
E	-12.507084	-12.507079	-12.507066	-12.507080
Cycles	7	22	10	2
Rotated				
r	1.0382	1.0381	1.0370	1.0381
θ	120°	120°	120°	120°
E	-12.507084	-12.507079	-12.507070	-12.507080
Cycles	7	26	12	2

and products towards the TS, but it also eliminates the problem associated with translations and rotations. On the other hand, the TS found by the App-LTP technique (Table 6.4) gives quite a good answer, and its optimization (App-LTP-GOPT), which requires only two more energy evaluations gives a final answer as good as does the AH technique. This outcome is in agreement with the two cases studied before.

Hydrogen Cyanide: HCN → CNH

Figure 6.2 shows a scheme of the HCN proton transfer reaction. The initial reactants, products and TS geometries, as well as a detailed LTP cycle-by-cycle uphill walk, which includes energy, geometry, and a number of update iterations to update the Hessian is summarized in Table 6.3.

Reactants   *Products*

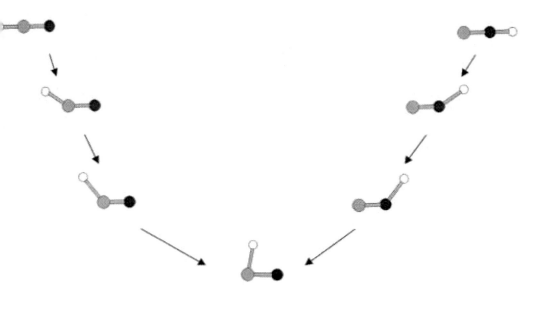

Transition State

Figure 6.2.   Proton transfer reaction of Hydrogen Cyanide. The behavior of the LTP technique is shown as reactants and products experience structural modifications, to coalesce finally to the TS with a H-C-N angle of 83 °.

A comparison of our results with those of other research groups is collected in Table 6.4. Zerner et.al. used the Augmented Hessian (AH) technique [20], but no bond lengths were reported. Bell and Crighton [14] on the other hand, have used an INDO Hamiltonian for their study, which has a different parametrization than the one contained in ZINDO. Their results are included here for geometry comparison purposes only.

In relation to the geometry, the angular coordinate obtained by Zerner et.al. [22] is reproduced by LTP, the difference being only 0.009° . There is, however, a difference of 0.038 Å for the H-C bond length and of 3.86° for the HCN angle between our results and those of Bell and Crighton, which we associate with the different parametrization of the Hamiltonians. The energy difference between AH and this work is of 0.3 kcal/mol.

In order to obtain further insight on the origin of these differences (that are, however, very small), we have performed a geometry optimization of the TS found by LTP, which required only 2 more SCF iterations. As a result, the

Table 6.3. LTP geometry, energy and number of update iterations (UI) required to find the TS for the hydrogen cyanide isomerization reaction (HCN → CNH). The results are those coming from the Exact Line-Then-Plane technique. The Energy (in Hartrees), the optimized geometry in a plane perpendicular to the walk direction, namely the hydrogen-carbon, the carbon-hydrogen, and the hydrogen-nitrogen bond lengths (in Angstroms) are shown. Also displayed are the hydrogen-carbon-nitrogen and the hydrogen-nitrogen-carbon angles (θ and φ, respectively) (in degrees). Finally, the number of update iterations of the Hessian for each cycle are displayed. The convergence criterion requires that the maximum component of the gradient should be smaller than 10^{-3} (Hartrees/Angstroms), whereas the search was started at an angle of $\theta = 130°$. The energy and geometry of the steps in the line are not shown, but the total number of energy evaluations was 32.

	Energy	r_{HC}	r_{CN}	r_{HN}	θ_{HCN}	φ_{HNC}	UI
R_o	-17.354816	1.0752	1.1840		130.00°		0
P_o	-17.313130		1.1991	1.0376		180.00°	0
R_1	-17.304199	1.0766	1.2161		104.65°		4
P_1	-17.312384		1.2024	1.0521		175.57°	4
R_2	-17.283882	1.1226	1.2285		77.78°		4
P_2	-17.291977		1.2041	1.0074		144.09°	3
R_3	-17.283736	1.0953	1.2195		88.72°		3
R_4	-17.282221	1.1039	1.2244		82.10°		2
P_4	-17.282171	1.1039	1.2232		84.31°		2
R_5	-17.282200	1.1092	1.2247		82.23°		2
P_5	-17.282156	1.1043	1.2233		84.16°		2

geometry improved but not significantly. Whereas the difference in the H-C bond length decreased only 0.005 Å, the CN bond length increased in 0.002 Å and the HCN angle by 0.22°. On the other hand, the decrease in energy accounts only for 0.3 kcal/mol. The number of iterations was not comparable because all the calculations have different starting geometries for the HCN angle α_o as shown in Table 6.4. However, it is interesting to notice that the number of SCF calculations was still in the range of procedures associated with the use of second derivatives.

From Table 6.3 we notice that the TS is located by the second iteration. It is in the reactant zone, with an HCN angle of 77.78°. From this point, the LTP algorithm resets the coordinates and searches in a smaller region of the hypersurface, accelerating convergence. At the next iteration (3) a test is performed at the reactants zone. The TS is again located in a narrower region of the space, and once more the coordinates are reset accordingly. A few final iterations are performed and the TS is found.

Table 6.4.   Geometry, energy, and number of SCF iterations required for the convergence to the TS for the HCN isomerization reaction. Displayed are the results obtained using the Augmented Hessian (AH) technique [20], Bell-Crighton's paper [1], and the LTP techniques. Also displayed on the last column is the TS found when an extra optimization calculation was performed on the LTP transition state already found. r_{CN} and r_{HC} are the corresponding carbon-nitrogen and hydrogen-carbon bond lengths (in Angstroms). θ is the hydrogen-carbon-nitrogen angle (in degrees), and E is the Energy (in Hartrees). For this technique, Zerner et al. [18] did not report the full geometry of the TS found. Their starting hydrogen-carbon-nitrogen angle was of 180°. For their work, Bell and Crighton used an INDO Hamiltonian that is different from the one used by Zerner and his coworkers., as well as by LTP. However, we include them here to compare the founded TS geometries. Their starting hydrogen-carbon-nitrogen angle was of 90°. They did not report the energetics of their calculations. The LTP starting hydrogen-carbon-nitrogen angle was $\theta = 160$?

	AH[a]	Bell Crighton[b]	LTP[c]	GOPT on LTP
r_{CN}		1.2254	1.2255	1.2239
r_{HC}		1.1397	1.1019	1.1073
θ	83°	79.23°	83.09°	83.22°
E	-17.282563		-17.282074	-17.282125
Iterations	15	11	13	2

Formic Acid

In Figure 6.3 the [1,3] Sigmatropic reaction of Formic Acid under study is shown. The reaction is characterized by a migration of the hydrogen atom with its sigma bond in a π carboxylic, i.e., the migration occurs by a shift in the π bonds of this metanoic acid environment.

The energy and geometry evolution of the LTP TS search technique for this reaction is shown in Table 6.5. The TS is characterized by the migrating proton bonded to both, the source and the migration oxygen atoms, by a oxygen-hydrogen bond length of $r_{24} = 1.2045$Å. Both oxygen atoms form angles with the carbon and the hydrogen attached to the carbon of $\theta_{215} = \theta_{315} = 129.82°$. As expected, both oxygen carbon bonds have now the same length, $r_{12} = r_{13} = 1.2960$ Å, and the oxygen-carbon-oxygen angle has narrowed by around 23°.

It is assumed that the mechanism of the reaction is a Symmetry-allowed Suprafacial sigmatropic shift reaction as reactants, products and TS are all in the same molecular plane.

Methyl Imine

This isomerization reaction occurs through 2 main competitive mechanisms shown in Figure 6.4. The first mechanism (a) involves motion, in the molecular

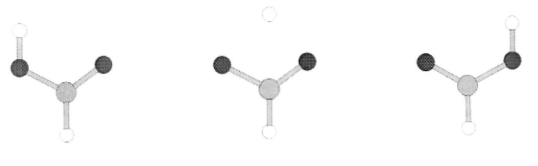

Reactants   *Transition State*   *Products*

Figure 6.3. The [1,3] sigmatropic reaction of formic acid. The initial reactants, the transition state found and the initial products structures are shown. Products are a mirror image of reactants. The geometrical details are given in Table 6.3.

Table 6.5. LTP geometry, energy, and number of iterations required to find the TS for the formic acid [1,3] Sigmatropic reaction. The results are those belonging to the Exact Line-Then-Plane technique. As reactants (R) and products (P) had the same geometry and energy we group them as: $Q_i = R_i = P_i$ for $i = 0, 1, 2, 3, 4$. The optimized geometry (bond length r are in Angstroms and angles θ are in degrees), in a plane perpendicular to the walking direction at each step, the energy (in kcal/mol) and the number of update iterations (UI) of the Hessian at each LTP cycle are displayed. The convergence criterion requires that the maximum component of the gradient and the distance be smaller than 10^{-3} (Hartrees/Angstroms). The energy and geometry of the steps in the line are not shown, but the total number of energy evaluations was 37. The labeling of the atoms is as shown in Figure 6.3.

	Energy*	r_{12}	r_{13}	r_{24}	θ_{213}	θ_{215}	θ_{315}	θ_{124}	UI
Q_0	0.0000	1.3298	1.2451	0.9976	123.15°	112.75°	124.10°	113.17°	0
Q_1	15.1640	1.3304	1.2527	1.0238	110.01°	123.37°	126.63°	89.93°	4
Q_2	32.7422	1.3157	1.2883	1.0764	104.82°	128.68°	126.50°	82.72°	3
Q_3	38.6872	1.3414	1.2421	1.1181	105.33°	126.19°	128.48°	79.18°	3
Q_4	47.7016	1.3084	1.2917	1.1656	100.55°	130.20°	129.26°	76.60°	2
R_5	49.2611	1.2960	1.2960	1.2045	100.37°	129.82°	129.82°	74.08°	2

*The origin of the energy (at Q_0) is: E = -41.394497 a.u.

plane, of hydrogen 5 (H_5) attached to the nitrogen passing through a planar C_{2v} conformation. The results for the LTP search for this mechanism are shown in Table 6.6, indicating that the TS is found at the 7th cycle, in the reactants zone with a hydrogen-nitrogen-carbon angle $\theta_{521} = 180°$ as expected. The other

significative change in the geometry is that the hydrogen-nitrogen bond length
is shortened during the reaction as seen in Table 6.6.

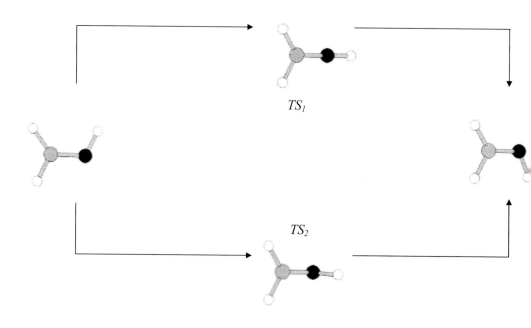

Reactants   Transition State   Product.

Figure 6.4. The methyl imine isomerization reaction. The initial reactants, the 2 possible
transition states and the initial products structures are shown for the 2 pathways for this reaction.
a) First transition state (TS$_1$) found by the Exact LTP technique. **b)** A second transition state
(TS$_2$) due to the internal rotation of the double bond is found. Products are a mirror image of
reactants. The geometric details and results for both mechanisms are shown in Tables 6.6 and
6.7, respectively.

The second mechanism (b) involves the internal rotation of the imine double
bond. This mechanism is performed through a reaction coordinate chosen to
be the dihedral angle α, as shown in Figure 6.4. This mechanism occurs when
the hydrogen (H$_5$) connected to the nitrogen comes out of the molecular plane
passing through a maxima at $\alpha = 90°$. The results for this mechanism have
been fitted through an interpolation procedure [28] and are represented here
in terms of the dihedral angle α, used as the reaction coordinate. The results
for this mechanism are shown in Table 6.7, in which only the most relevant
geometrical changes have been displayed. The energy difference between these
two possible mechanisms indicates that mechanism (a) appears to be favored
over mechanism (b).

Table 6.6. LTP geometry, energy, and number of iterations required to find the TS for the methyl imine isomerization reaction. The results are those belonging to the Exact Line-Then-Plane technique. As reactants (R) and products (P) had the same energy and almost the same geometry, we group them as: $Q_i = R_i = P_i$ for $i = 0, 1, 2, 3, 4$. The optimized geometry (bond length r are in Angstroms and angles θ are in degrees), in a plane perpendicular to the walking direction at each step, the energy (in kcal/mol) and the number of update iterations (UI) of the Hessian at each LTP cycle are displayed. The convergence criterion requires that the maximum component of the gradient be smaller than 10^{-3}. The energy and geometry of the steps in the line are not shown, but the total number of LTP cycles was 73. The labeling of the atoms is as shown in Figure 6.4.

	Energy[a]	r_{12}	r_{13}	r_{14}	r_{25}	θ_{213}	θ_{215}	θ_{315}	θ_{124}	UI
Q_0	0.0000	1.2884	1.0941	1.0941	1.0546	122.38	122.38	115.24	112.72	0
Q_1	5.5596	1.2812	1.0870	1.1019	1.0323	121.79	123.02	115.2	131.08	6
Q_2	12.6725	1.2766	1.1025	1.0883	1.0291	123.07	121.90	115.04	142.77	4
Q_3	18.1732	1.2737	1.1029	1.0895	1.0289	123.09	122.00	114.91	141.05	5
Q_4	22.0346	1.2714	1.1027	1.0910	1.0291	123.08	122.13	114.80	157.22	5
Q_5^b	29.1952	1.2679	1.0996	1.0969	1.0299	122.83	122.61	114.56	175.56	6
Q_6^c	29.4964	1.2679	1.0983	1.0978	1.0297	122.71	122.7	114.59	179.11	3
R_5	29.5083	1.2679	1.0981	1.0980	1.0297	122.71	122.70	114.59	179.82	0

[a] The origin of the energy (at Q_0) is: $E = -18.888477$ a.u.
[b] At this cycle the step factor is reseted to 2.5, according to the LTP algorithm.
[c] At this cycle the step factor is amplified by a factor of 2, according to the LTP algorithm.

Table 6.7. Internal rotation of methyl imine according to mechanism (b) of Figure 6.4. The TS is found at a dihedral angle of $\alpha = 90$. The energy (in kcal/mol) and relevant geometrical parameters of the uphill search are shown. The labeling of the geometrical parameters is as shown in Figure 6.4.

α	Energy*	r_{25}	θ_{125}
$0° = 180°$	0.0000	1.0546	112.72°
$20° = 160°$	8.8409	1.1026	111.74°
$40° = 140°$	31.2268	1.2242	109.27°
$60° = 120°$	56.6831	1.3624	106.46°
$80° = 100°$	73.2985	1.4526	104.62°
$90°$	75.5774	1.4650	104.37°

*The origin of the energy (at P_0) is:
$E = -40.490650$ a.u.

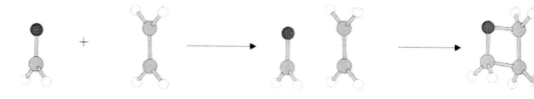

Reactants   Transition State   Products

Figure 6.5. The thermal retro [2+2] cycloaddition reaction of oxetane mechanism. The initial reactants, the transition state found and the initial products structure are shown. The geometrical details of the search are shown in Table 6.8.

Thermal Retro [2+2] Cycloaddition of Oxetane

The mechanism considers two molecules reacting to give products. The molecular planes of reactants are parallel to each other, and at a distance of 3.0 Å ($r_{27} = r_{18} = 3.0$ Å).

In Figure 6.5, the mechanism of the reaction is shown as well as the labeling of the atoms employed. The results of the LTP uphill search are shown in Table 6.8. The TS found shows an elongation of the ethane C_1-C_2 bond length, from 1.3238 Å at the initial reactants structure to 1.4082 Å at the TS, which is midway between the bond of ethene and the single oxo-cycle bond. The same tendency is observed for the C_7-O_8. These changes are expected as the new single bonds C_1-C_7 and C_2-O_8 are coming to a distance closer to a single bond, while the original double bonds migrate to intermediate single-double bond distances.

Our UHF results allow us to conclude that the reaction occurs in a concerted fashion, as no intermediate of the reaction was found.

5.   Concluding remarks

The transition state search results using the Line-Then-Plane technique presented in this work, suggest that for a model to be successful and expedite in the search for TS, both reactant and product should be considered. This concept of considering **R**'s and **P**'s approaching the TS simultaneously from both sides of the reaction is mainly based on the idea that each structure contains information and history about the other, having in common the same TS. The trajectory drawn in this way can be thought of as the lowest passage from one side of a mountain (**R** or **P**) to the other (**P** or **R**). The advantage of this idea is that we look at the mountain from both sides, and not from one only.

One of the main features of LTP is that it does not require a guess of the TS, nor does it need an evaluation of the Hessian. The Hammond Adapted LTP (HALTP) procedures seem to require a small number of calculations, but they still find the TS with the same accuracy. Although these procedures are simple,

Table 6.8. LTP energy (kcal/mol), geometry (bond lengths r are in Angstroms and angles θ are in degrees) and the number of update iterations (UI) required to find the TS for the thermal retro [2+2] cycloaddition reaction of Oxetane. The results belong to the Exact LTP technique. The convergence criteria were that the maximum component of the gradient and the displacement vector should be smaller than 10^{-3} Hartrees/Angstroms. The energy and geometry of the steps in the line are not shown. The total number of LTP cycles was 47. The labeling of the atoms is as shown in Figure 6.5.

	Energy^a	$|d|$	r_{12}	r_{78}	r_{17}	r_{28}	θ_{287}	UI
R_o	336.6424	3.1202	1.3238	1.2366	3.0000	3.0000		0
P_o	0.0000		1.5007	1.3849	1.5000	1.3842	88.65°	0
R_1	337.6691	2.4962	1.3403	1.2512				2
P_1	49.9408		1.4819	1.3699	1.6498	1.5459	88.87°	2
R_2	340.6801	1.9969	1.3537	1.2629				2
P_2	135.2269		1.4671	1.3579	1.7697	1.6753	89.07°	2
R_3	344.3734	1.5976	1.3646	1.2723				2
P_3	215.0582		1.4553	1.3483		1.7788	89.23°	2
R_4	348.0924	1.2781	1.3733	1.2799				2
P_4	280.1598		1.4459	1.3345				2
R_5	351.5258	1.0224	1.3804	1.2859				3
P_5	329.8571		1.4385	1.3345				2
R_6	354.5469	0.8178	1.3861	1.2907				2
P_6	364.1506		1.4325	1.3296				2
R_7^b	365.5244	0.1636	1.4045	1.3063				2
P_7	371.1769		1.4138	1.3140				2
R_8^c	367.8363		1.4082	1.3094				0

^{a} The origin of the energy (at P_0) is: E = -40.490650 a.u.
^{b} At this cycle the step factor is reseted to 2.5, according to the LTP algorithm.
^{c} At this cycle the TS is found in the Reactants zone. LTP stops.

they have the advantage of using a reduced number of calculations, a simple and convenient way to write the projected coordinates, and update techniques for the Hessian. It considers the existence of an intermediate of reaction and embraces the idea of finding the TS(s) starting simultaneously from **R**'s and **P**'s.

The procedures studied in this work may fail in cases characterized by a very steep reaction path. In those cases, it appears to be necessary to reconstruct more of the reaction path to ensure that the TS has been found, and to reproduce the reaction path more accurately. This can be pursued by connecting the first-order saddle point with reactants and products by means of a down-hill procedure, such as the steepest descent technique (despite the inherent extra cost), especially if one is interested in kinetic aspects of the reaction. However, lower energy saddle points may exist.

Finally, it has become evident over the years, and according to the large amount of work devoted to tackling the TS search problem, that chemical intuition is always required in this kind of problems. Unfortunately, if it fails one is lost in the hypersurface of algorithms with no saving recipe available.

Acknowledgments

Cristian Cardenas-Lailhacar would like to dedicate this work to the memory of his advisor, mentor, professor, and friend, the Distinguished Professor of Chemistry and Physics, Dr. Michael Charles Zerner. Helpful discussions with Dr. Keith Runge from the Quantum Theory Project are gratefully acknowledged. Computational time provided by the Quantum Theory Project, at the University of Florida is also acknowledged. This work was supported in part by the U.S. Office of Naval Research.

References

[1] Th. A. Halgren and W. N. Lipscomb, Chem. Phys. Lett. 49 (2), 225 (1977).

[2] A. Jensen, Theor. Chim. Acta 63, 269 (1983).

[3] Ch. Cerjan and W. H. Miller, J. Chem. Phys. 75 (6), 2800 (1981).

[4] J. Simons, P. Jorgensen, H. Taylor and J. Ozment, J. Phys. Chem. 87, 2745 (1983).

[5] R. Fletcher and M. J. D. Powell, Comput. J. 6, 163 (1963) ; W.C. Davidon, Comput. J. 10, 406 (1968); R. Fletcher, Comput. J. 8, 33 (1965).

[6] A. Banerjee, N. Adams, J. Simons and R. Shepard, J. Phys. Chem. 89, 52 (1985).

[7] D. K. Hoffman, R. S. Nord and K. Ruedenberg, Theor. Chim. Acta 69, 265 (1986). See also: M. V. Basilevsky, Chem. Phys. 67, 337 (1982).

[8] H. B. Schlegel, Theor. Chim. Acta 83, 15 (1992).

[9] P. Culot, G. Dive, V. H. Nguyen and J. M. Ghuysen, Theor. Chim. Acta 82, 189 (1992).

[10] K. Ruedenberg and J-Q. Sun, J. Chem. Phys., 100, 5836 (1994); J-Q. Sun and K. Ruedenberg, J. Chem. Phys., 101 (3), 2157 (1994); J-Q. Sun and K. Ruedenberg, J. Chem. Phys., 101 (3), 2168 (1994).

[11] C. Gonzales and B. Schlegel, J. Phys. Chem. 95 (8), 5853 (1991).

[12] C. Gonzalez and H. B. Schlegel, J. Chem. Phys. 90, 2154 (1989), C. Gonzalez and H. B. Schlegel, J. Phys. Chem. 94, 5523 (1990).

[13] Y. Abashkin and N. Russo, J. Chem. Phys. 100 (6), 4477 (1994).

[14] S. Bell and J. Crighton, J. Chem. Phys. 80 (6), 2464 (1984).

[15] K. M. Dunn and K. Morokuma, J. Chem. Phys. 102, 4904 (1995).

[16] M. J. S. Dewar, E. F. Healy and J. J. P. Stewart, J. Chem. Soc., Faraday Trans. 11 80, 227 (1984); M. J. S. Dewar, E. G. Zoebisch, E. F. Healy and J. J. P. Stewart, J. Am. Chem. Soc.107, 3902 (1985).

[17] C. Cardenas-Lailhacar and M.C. Zerner, Int. J. Quantum Chem. 55, 429 (1995).

[18] M. C. Zerner, Department of Chemistry, University of Florida, Gainesville, Florida 32611. ZINDO, A General Semi-Empirical Program Package.

[19] C. G. Broyden, Math. Comput. 21, 368 (1967); R. Fletcher, Comput. J. 13, 23 (1970); Goldfarb, Math. Comput. 24, 23 (1970): D. F. Shanno, Math. Comput. 24, 647 (1970).

[20] B.H. Lengsfield, J. Chem. Phys. 73, 382 (1980)

[21] M. C. Zerner, in Appendix C of A. Szabo and N. S. Ostlund, *Modem Quantum Chemistry,* Mc Graw-Hill, New York, USA, 1989.

[22] J. D. Head, B. Weiner and M. C. Zerner, Int. J. Quantum. Chem. 33, 177 (1988).

[23] J. D. Head and M. C. Zerner, Chem. Phys. Lett. 122, 264 (1985).

[24] J. D. Head and M. C. Zerner, in: Geometrical Derivatives of Energy Surfaces and Molecular Properties, Eds. P. Jorgensen and J. Simons (Reidel, Dordrecht, 1985).

[25] J. D. Head and M. C. Zerner, Chem. Phys. Lett. 131, 359 (1986).

[26] R. Fletcher, *Practical Methods of Optimization,* Wiley, Chichester, 1981.

[27] J. W. McIver and A. Komornicki, J. Am. Chem. Soc. 94, 2625 (1972).

[28] C. Cardenas-Lailhacar and M.C. Zerner, Int. J. Quantum Chem. 75, 563 (1999).

Chapter 7

A SIMPLE APPROXIMATION ALGORITHM FOR NONOVERLAPPING LOCAL ALIGNMENTS (WEIGHTED INDEPENDENT SETS OF AXIS PARALLEL RECTANGLES)

Piotr Berman
Department of Computer Science & Engineering
Pennsylvania State University
University Park, PA 16802
berman@cse.psu.edu

Bhaskar DasGupta[*]
Department of Computer Science
Rutgers University
Camden, NJ 08102
bhaskar@crab.rutgers.edu

Abstract   We consider the following problem motivated by applications to nonoverlapping local alignment problems in computational molecular biology: we are a given a set of n positively weighted axis parallel rectangles such that, for each axis, the projection of a rectangle on this axis does not enclose that of another, and our goal is to select a subset of *independent* rectangles from the given set of rectangles of total maximum weight, where two rectangles are independent provided for each axis, the projection of one rectangle does not overlap that of another. We use the two-phase technique of [3] to provide a simple approximation algorithm for this problem that runs in $O(n \log n)$ time with a worst-case performance ratio of 3. We also discuss extension and analysis of the algorithm in d dimensions.

Keywords: Computational Biology, Nonoverlapping Local Alignments, Approximation Algorithms.

---

[*]Supported in part by NSF grant CCR-9800086.

P.M. Pardalos and J. Principe (eds.), Biocomputing, 129-138.

1.   Introduction

A fundamental problem that arises in the comparison of genomic sequences in computational molecular biology for similarity or dissimilarity is to select fragments of high local similarity between two strings [6]. Motivated by this application, Bafna et al. [1], considered the following maximization problem, termed as the Independent subset of Rectangles (IR) problem. We are a given a set S of n positively weighted axis parallel rectangles such that, for each axis, the projection of a rectangle on this axis does not enclose that of another. Define two rectangles to be independent if for each axis, the projection of one rectangle does not overlap that of another. The goal of the IR problem is to select a subset $S' \subseteq S$ of *independent* rectangles from the given set of rectangles of total maximum weight. The *unweighted* version of the IR problem is the one in which the weights of all rectangles are identical. See Figure 7.1 for an pictorial illustration of the problem. The reader is referred to Section 2 of [1] for a detailed description of the relationship of this problem to the local alignment methods; Figure 7.2 shows a pictorial illustration of the relationship of a rectangle to local similarity between two fragments of two sequences.

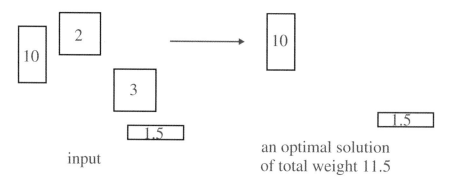

Figure 7.1.   An illustration of the IR problem

A summary of previous results on this problem is as follows. Halldórsson [5] provided a polynomial time approximation algorithm with a performance ratio of $2 + \varepsilon$ (for any constant $\varepsilon > 0$) for the unweighted version of the IR problem[1]. Bafna et al. [1] showed that an approach similar to that in [5] yields a polynomial time approximation algorithm with a performance ratio of $\frac{13}{4}$ for the IR problem. The current best approximation algorithm for the IR problem is due to Berman [2] which has a performance ratio of $\frac{5}{2} + \varepsilon$ (for any constant $\varepsilon > 0$).

Consider the graph G formed from the given rectangles in which there is a node for every rectangle with its weight being the same as that of the rectangle and two nodes are connected by an edge if and only if their rectangles are *not*

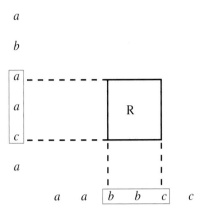

Figure 7.2. The rectangle R captures the local similarity (match) between the fragments aac and bbc of the two sequences; weight of R is the strength of the match.

independent. It is not difficult to see that G is a 5-claw free graph [1] and the IR problem is tantamount to finding a *maximum-weight* independent set in G. Previous approaches have used this connection of the IR problem to the 5-claw free graphs to provide better approximation algorithms by giving improved approximation algorithms for d-claw free graphs. Most of these algorithms essentially start with an arbitrary solution and then allows small improvements to enhance the approximation quality of the solution. In contrast, we consider the IR problem directly and use the simple greedy two-phase technique of [3] to provide an approximation algorithm for the IR problem that runs in $O(n \log n)$ time with a performance ratio of 3. Although our approximation algorithm does not improve the worst-case performance ratios of previously best algorithms, it is simple to implement (involving standard simple data structures such as stacks and binary trees) and runs faster than the algorithms in [1, 2, 5].

The following notations and terminologies are used for the rest of this paper. An interval $[a, b]$ is the set $[a, b] = \{x \in \mathbb{R} : a \leq x \leq b\}$. A rectangle R is $[a, b] \times [c, d]$ for some two intervals $[a, b]$ and $[c, d]$, where $\times$ denotes the Cartesian product. The weight of a rectangle R is denoted by $w(R)$. The *performance ratio* of an approximation algorithm for the IR problem is the ratio of the total weights of rectangles in an optimal solution to that in the solution provided by the approximation algorithm. We assume that the reader with familiar with standard techniques and data structures for the design and analysis of algorithms such as in [4].

2. Application of the Two-Phase technique to the IR problem

Let $R_1, R_2, \ldots, R_n$ be the n input rectangles in our collection, where $R_i = X_i \times Y_i$ for some two intervals $X_i = [d_i, e_i]$ and $Y_i = [f_i, g_i]$. Consider the intervals $X_1, X_2, \ldots, X_n$ formed by projecting the rectangles on one axis and call two intervals X_i and X_j independent if and only if the corresponding rectangles R_i and R_j are independent. The notation $X_i \simeq X_j$ (respectively, $X_i \not\simeq X_j$) is used to denote if two intervals X_i and X_j are independent (respectively, not independent).

To simplify implementation, we first sort the set of numbers $\{d_i, e_i \mid 1 \leq i \leq n\}$ (respectively, the set of numbers $\{f_i, g_i \mid 1 \leq i \leq n\}$) and replace each number in the set by its rank in the sorted list. This does not change any feasible solution to the given problem; however, after this $O(n \log n)$ time preprocessing we can assume that $d_i, e_i, f_i, g_i \in \{1, 2, \ldots, 2n\}$ for all i. This assumption simplifies the design of data structures for the IR problem.

Now, we adopt the two-phase technique of [3] on the intervals $X_1, X_2, \ldots, X_n$. The precise algorithm is shown below in Figure 7.3. The solution to the IR problem consists of those rectangles whose projections are returned in the solution at the end of the selection phase.

The main theorem of this section is as follows.

Theorem 1 *Algorithm TPA-IR provides a correct solution to the IR problem in* $O(n \log n)$ *time with a performance ratio of 3.*

Before proving the performance guarantees of Theorem 1, we will first prove the correctness and running time of Algorithm TPA-IR.

Lemma 2 *Algorithm TPA-IR returns a correct solution to the IR problem in* $O(n \log n)$ *time.*

Proof. To show that the algorithm is correct we just need to show that the selected rectangles are mutually independent. This is obviously ensured by the final selection phase.

It takes $O(n \log n)$ time to create the list **L** by sorting the endpoints of the n rectangles. It is easy to see that, for each interval $X_i = [d_i, e_i]$ $(1 \leq i \leq n)$, the algorithm performs only a constant number of operations, which are elementary except for the computation of TOTAL(X_i) in the evaluation phase. We need to show how this function can be computed in $O(\log n)$ time for each X_i. Note that $X_i \not\simeq X_j \equiv (X_i \cap X_j \neq \emptyset) \vee (Y_i \cap Y_j \neq \emptyset)$. Since the intervals are considered in non-decreasing order of their endpoints, there is no interval in **S** with an endpoint to the right of e_i when TOTAL(X_i) is computed. As a result, when the computation of TOTAL(X_i) is needed, $X_i \not\simeq X_j$ for an X_j currently in stack provided *exactly* one of the following two conditions is satisfied:

(a) $e_j \geq d_i$,

(b) $(e_j < d_i) \wedge (Y_i \cap Y_j \neq \emptyset)$.

Since for any two intervals $Y_i = [f_i, g_i]$ and $Y_j = [f_j, g_j]$, such that neither interval encloses the other, $[f_i, g_i] \wedge [f_j, g_j] \neq \emptyset$ is equivalent to either $f_i \leq f_j \leq$

(* definitions *)
    a *triplet* (α, β, γ) is an ordered sequence of
        three values α, β and γ;
    L is sequence that contains a triplet $(w(R_i), d_i, e_i)$
        for every $R_i = X_i \times Y_i$ with $X_i = [d_i, e_i]$;
        L is sorted so the values of e_i's are in non-decreasing order;
    S is an initially empty stack that stores triplets;
    TOTAL(X_j) returns the sum of v's of those triplets (v, a, b) $\in$**S**
        such that $[a, b] \not\simeq X_j$;

(* evaluation phase *)
    for (each $(w(R_i), d_i, e_i)$ from **L**)
    {
        $v \leftarrow w(R_i) - $ TOTAL$([d_i, e_i])$;
        if ($v > 0$)
            push$((v, d_i, e_i),$**S**$)$;
    }

(* selection phase *)
    while (**S** is not empty)
    {
        $(v, d_i, e_i) \leftarrow$ pop(**S**);
        if ($[d_i, e_i] \simeq X$ for every interval X in our solution)
            insert $[d_i, e_i]$ to our solution;
    }

Figure 7.3. Algorithm TPA-IR: Adoption of the two-phase technique for the IR problem.

g_i or $f_i \leq g_j \leq g_i$ but not both, it follows that for the purpose of computing
TOTAL(X_i) it suffices to maintain a a data structure $\mathcal{D}$ for a set of points in the
plane with coordinates from the set $\{1, 2, \ldots, 2n\}$ such that the following two
operations can be performed:

Insert(v, x, y): Insert the point with coordinates (x, y) (with $x, y \in \{1, 2, \ldots, 2n\}$)
and value v in $\mathcal{D}$. Moreover, if Insert(v, x, y) precedes Insert(v', x', y'),
then $y' \geq y$.

Query(a, b, c): Given a query range (a, b, c) (with $a, b, c \in \{1, 2, \ldots, 2n\} \cup$
$\{-\infty, \infty\}$), find the sum of the values of all points (x, y) in $\mathcal{D}$ with
$a \leq x \leq b$ and $y \geq c$.

For example, finding all X_j's currently in stack with $e_j \geq d_i$ is equivalent to doing Query$(-\infty, \infty, d_i)$.

For notational simplicity, assume that $n = 2^k$ for some positive integer k. We start with a skeleton rooted balanced binary tree T with $2n$ leaves (and of height $O(\log n)$) in which each node will store a number in $\{1, 2, \ldots, 2n\}$. The ith leaf of T (for $1 \leq i \leq 2n$) will store the point (i, y), if such a point was inserted. With each node v of T, we also maintain the following:

- a list L_v of the points stored in the leaves of the subtree rooted at v, sorted by their second coordinate. Additionally, each entry (x, y) in the list also has an additional field $sum_{x,y}$ storing the sum of all values of all points in the list to the left of (x, y) including itself, that is, the sum of values of all points (x', y') in L_v with $y' \leq y$.

- a value s_v equal to the sum of values of all points in L_v.

Initially $L_v = \emptyset$ and $s_v = 0$ for all $v \in T$ and building the skeleton tree thus obviously takes $O(n)$ time.

To implement Insert(v, x, y), we insert the point (x, y) at the xth leaf of T and update L_v and s_v for every node v on the unique path from the root to the xth leaf. Since Insert(v, x, y) precedes Insert(v', x', y') implies $y' \geq y$, we simply append (x, y) to the existing L_v for every such v. It is also trivial to update $sum_{x,y}$ (from the value of $sum_{x',y'}$ where (x', y') was the previous last entry of the list) and s_v in constant time. Since there are $O(\log n)$ nodes on the above unique path, we spend $O(\log n)$ time.

To implement Query(a, b, c), we search for a and b in T (based on the first coordinates of the points only) as in a binary search tree and let v be the lowest common ancestor of the two search paths. Then L_v contains all points (x, y) with $a \leq x \leq b$. We do a binary search on L_v in $O(\log |L_v|) = O(\log n)$ time based on the second coordinate of points to find a point (x', y') with y' being the largest possible value satisfying $y' < c$. Then, the answer to our query is the quantity $s_v - sum_{(x',y')}$. $\square$

Now, we prove the performance ratio of Algorithm TPA-IR as promised in Theorem 1. Let B be a solution returned by Algorithm TPA-IR and A be any optimal solution. For a rectangle $R \in A$, let β_R denote the number of those rectangles in B that were *not* independent of R *and* were examined no earlier than R by the evaluation phase of Algorithm TPA-IR and let $\beta = \max_{R \in A} \beta_R$.

Theorem 3 *Algorithm TPA-IR has a performance ratio of β.*

Proof. Consider the set of intervals $\mathbf{S}$ in the stack at the end of the evaluation phase. Let $W(A) = \sum_{R_i \in A} w(R_i)$ and $V(\mathbf{S}) = \sum_{(v,d_i,e_i) \in \mathbf{S}} v$. It was shown in Lemma 3 of [3] that the the sum of the weights of the rectangles selected during the selection phase is at least $V(\mathbf{S})$. Hence, it suffices to show that $\beta V(\mathbf{S}) \geq W(A)$.

Consider a rectangle $R_i = X_i \times Y_i \in A$ and the time when the evaluation phase starts the processing of $X_i = [d_i, e_i]$. Let $\text{TOTAL}'([d_i, e_i])$ and $\text{TOTAL}''([d_i, e_i])$ be the values of $\text{TOTAL}([d_i, e_i])$ before and after the processing of X_i, respectively.

If $w(R_i) < \text{TOTAL}'([d_i, e_i])$, then X_i is not pushed to the stack and $\text{TOTAL}''([d_i, e_i]) = \text{TOTAL}'([d_i, e_i])$. On the other hand, if $w(R_i) \geq \text{TOTAL}'([d_i, e_i])$, then X_i is pushed to the stack with a value of $w(R_i) - \text{TOTAL}'([d_i, e_i])$, as a result of which $\text{TOTAL}''([d_i, e_i])$ becomes at least $\text{TOTAL}'([d_i, e_i]) + (w(R_i) - \text{TOTAL}'([d_i, e_i])) = w(R_i)$. Hence, in either case $\text{TOTAL}''([d_i, e_i]) \geq w(R_i)$.

Now, summing up over all R_i's and using the definition of β and $\text{TOTAL}''([d_i, e_i])$ gives

$$W(A)$$

$$= \sum_{R_i \in A} w(R_i)$$

$$\leq \sum_{R_i = [d_i, e_i] \times Y_i \in A} \text{TOTAL}''([d_i, e_i])$$

$$\leq \sum_{((v,a,b) \in \mathbf{S}) \wedge (b \leq e_i) \wedge ([a,b] \not\supseteq [d_i, e_i])} v \qquad \text{(by definition of TOTAL}''([d_i, e_i]))$$

$$\leq \beta \sum_{(v,a,b) \in \mathbf{S}} v \qquad \text{(by definition of } \beta)$$

$$= \beta V(\mathbf{S})$$

❑

The proof of Theorem 1 can now be completed by proving the following Lemma.

Lemma 4 *For the IR problem, $\beta = 3$.*

Proof. First note that $\beta = 3$ is possible; see Figure 7.4.

Now we show that $\beta > 3$ is impossible. Refer to Figure 7.4. Remember that rectangles in an optimal solution contributing to β must not be independent of our rectangle R and must have their right vertical right on or to the right of the vertical line L. Since rectangles in an optimal solution must be independent of

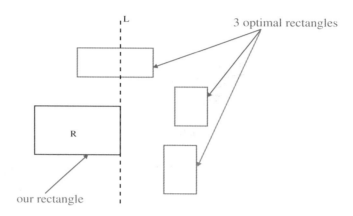

Figure 7.4. A tight example for Algorithm TPA-IR showing $\beta = 3$ is possible.

each other, there can be at most one optimal rectangle crossing L (and, thereby conflicting with R in its projections on the x-axis). Any other optimal rectangle must lie completely to the right of L and therefore may conflict with R in their projections on the y-axis only; hence there can be at most two such rectangles. ❑

3. Concluding remarks

Algorithm TPA-IR makes a pass on the projections of the rectangles on the x-axis in a nondecreasing order of the endpoints of the projections. Can we improve the performance ratio if we run TPA-IR separately on the projections on the x-axis in left-to-right and in right-to-left order of endpoints and take the better of the two solutions? Or, even further, we may try running Algorithm TPA-IR two more times separately on the projections on the y-axis in top-to-bottom and in bottom-to-top order and take the best of the four solutions. Figure 7.5 shows that even then the worst case performance ratio will be 3. We already exploited the planar geometry induced by the rectangles for the IR problem to show that $\beta \leq 3$. Further research may be necessary to see whether we can exploit the geometry of rectangles more to design simple approximation algorithms with performance ratios better than 2.5 in the weighted case or better than 2 in the unweighted case.

In practice, the d-dimensional variation of the IR problem, motivated by the selection of fragments of high local similarity between d strings, is more important. In this version, we are given a set S of n positively weighted axis parallel d-dimensional hyper-rectangles[2] such that, for every axis, the projection of a hyper-rectangle on this axis does not enclose that of another. Defining two hyper-rectangles to be independent if for every axis, the projection of one

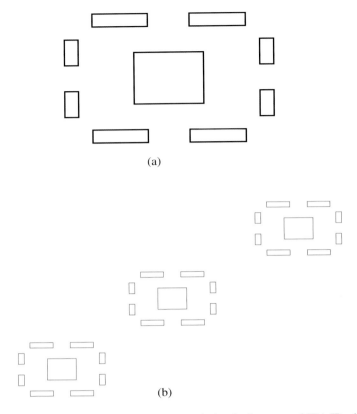

Figure 7.5. (a) The basic block of 9 rectangles such that the four runs of TPA-IR will always select R, whereas an optimal solution will select 3 of the remaining 8 rectangles. (b) The basic block repeated $\frac{n}{9}$ times such that different copies do not interfere in their projections on either axis, resulting in a performance ratio of 3 even for the best of the 4 runs.

hyper-rectangle does not overlap that of another, the goal of the d-dimensional IR problem is to select a subset $S' \subseteq S$ of *independent* hyper-rectangles from the given set of rectangles of total maximum weight. Algorithm TPA-IR can be applied in an obvious way to this extended version by considering the projections of these hyper-rectangles on a particular axis. It is not difficult to see that $\beta \le 2^d - 1$ for this case, thus giving a worst-case performance ratio of $2^d - 1$. Whether one can design an algorithm with a performance ratio that increases less drastically (e.g., linearly) with d is still open.

Notes

1. For this and other previous approximation algorithms with an ε in the performance ratio, the running time increases with decreasing ε, thereby rendering these algorithms impractical if ε is small. Also, a straightforward implementation of these algorithms will run in at least $\Omega(n^2)$ time.

2. A d-dimensional hyper-rectangle is a Cartesian product of d intervals.

References

[1] V. Bafna, B. Narayanan and R. Ravi, *Nonoverlapping local alignments (Weighted independent sets of axis-parallel rectangles*, Discrete Applied Mathematics, 71, pp. 41-53, 1996.

[2] P. Berman, *A $d/2$ approximation for maximum weight independent set in d-claw free graphs*, proceedings of the 7th Scandinavian Workshop on Algorithmic Theory, Lecture Notes in Computer Science, 1851, Springer-Verlag, July 2000, pp. 214-219.

[3] P. Berman and B. DasGupta, *Improvements in Throughput Maximization for Real-Time Scheduling*, proceedings of the 32nd Annual ACM Symposium on Theory of Computing, May 2000, pp. 680-687.

[4] T. H. Cormen, C. E. Leiserson and R. L. Rivest, Introduction to Algorithms, The MIT Press, 1990.

[5] M. M. Halldórsson, *Approximating discrete collections via local improvements*, proceedings of the 6th ACM-SIAM Symposium on Discrete Algorithms, January 1995, pp. 160-169.

[6] T. F. Smith and M. S. Waterman, *The identification of common molecular sequences*, Journal of Molecular Biology, 147, 1981, pp. 195-197.

Chapter 8

COMBINED APPLICATION OF GLOBAL OPTIMIZATION AND NONLINEAR DYNAMICS TO DETECT STATE RESETTING IN HUMAN EPILEPSY *

J.C. Sackellares
Neurology; Bioengineering; Neuroscience
University of Florida
sackellares@epilepsy.health.ufl.edu

L.D. Iasemidis
Bioengineering
Center for Systems Science and Engineering Research
Arizona State University
leon.iasemidis@asu.edu

P. Pardalos
Center for Applied Optimization
Industrial and Systems Engineering
University of Florida
pardalos@ufl.edu

D.-S. Shiau
Statistics
University of Florida
shiau@epilepsy.health.ufl.edu

Abstract   Epilepsy is a common neurological disorder characterized by recurrent seizures, most of which appear to occur spontaneously. Our research, employing novel

---

*This research is supported by NIH, NSF, VA, Whitaker and DARPA research grants.

P.M. Pardalos and J. Principe (eds.), Biocomputing, 139-157.

signal processing techniques based on the theory of nonlinear dynamics, led us to the hypothesis that seizures represent a spatiotemporal state transition in a complex chaotic system. Through the analysis of long-term intracranial EEG recordings obtained in patients with medically intractable seizures, we discovered that seizures were preceded by a preictal transition that evolves over tens of minutes. This transition is followed by a seizure. Following the seizure, the spatiotemporal dynamics appear to be reset. The study of this process has been hampered by its complexity and variability. A major problem was that the transitions involve a subset of brain sites that vary from seizure to seizure, even in the same patient. However, by combining dynamical analytic techniques with a powerful global optimization algorithm for selecting critical electrode sites, we have been able to elucidate important dynamical characteristics underlying human epilepsy. We illustrate the use of these approaches in confirming our hypothesis regarding postictal resetting of the preictal transition by the seizure. It is anticipated that these observations will lead to a better understanding of the physiological processes involved. From a practical perspective, this study indicates that it may be possible to develop novel therapeutic approaches involving carefully timed interventions and reset the preictal transition of the brain well prior to the onset of the seizure.

1.   Introduction

Epilepsy is a common neurological disorder affecting approximately 1 out of every 100 people of all ages. It is characterized by recurrent seizures which cause transient neurological disturbances ranging from strange feelings to involuntary motor activity to impaired cognitive processes to complete loss of consciousness. Seizures typically last for 1 to 5 minutes. They are often followed by a post-seizure (postictal) period characterized by confusion, disorientation, drowsiness, and fatigue. The postictal symptoms can last for minutes to hours. Clinical seizures result from the sudden onset of a highly ordered discharge of millions of neurons in the cerebrum of the brain. The exact location and pattern of these seizure discharges determine the clinical character of the seizure.

Most research into the physiological mechanisms that lead to epileptic seizures has focused upon the neuronal processes that occur during seizures or during brief interictal discharges (spikes) that traditionally were seen as larval seizures. In our research, we have sought to understand the mechanisms leading to seizures by analyzing long-term electroencephalographic (EEG) recordings, of many days duration, that include many epileptic seizures. Rather than focus exclusively upon the seizure itself, we have extended our analysis to the periods before and after recorded seizures. Our objective has been to characterize the dynamics of the transition from the asymptomatic interictal state to the seizure, through the postictal state, then back to the interictal state. We have conceptualized these physiological transitions as state transitions of the brain. Because these transitions usually occur without any observable external trigger, it seems likely that the transitions are due to the dynamical characteristics of the epileptic brain itself. Given the complexity of the brain with its millions of feedback loops (excitatory and inhibitory), we postulated that the brain is a highly complex chaotic system. Thus, we developed signal processing techniques that are based upon the theories of nonlinear dynamics and chaos [9, 7].

It is our hypothesis that seizures are an example of the intermittent self organizing phase transitions that occur in chaotic systems. The epileptogenic focus appears to recruit other more normal areas of the cortex into a pathological preictal dynamical state that leads to a seizure [9, 7]. Further, we postulate that the seizure serves to reset the system [22, 23, 14].

Because of the limitations of traditional signal processing techniques, we had to develop novel EEG signal processing algorithms based upon the theory of nonlinear dynamics [9, 7]. Using these techniques, we analyzed long-term, continuous multichannel intracranial EEG recordings obtained for clinical diagnostic purposes in patients with temporal and frontal lobe epilepsy. By calculating the value of the largest Lyapunov exponent for sequential 10.24 second epochs of EEG signals recorded from each of the multiple intracranial electrode channels, we were able to characterize important spatiotemporal dynamical features of the epileptic brain. An example of a typical intracranial electrode pattern is shown in figure 1. We subsequently defined the sequential Lyapunov measures as the short-term Lyapunov exponent (STL_{max}). This measure provides a quantitative description of the magnitude of order or chaos present in the EEG signal at a given segment in time from each electrode site.

Our first observation was that, for many sites of the epileptogenic cerebral cortex, STL_{max} was relatively high prior to a seizure (preictal period). During the seizure (ictal period), there is an abrupt drop in the value of STL_{max}. After a seizure, STL_{max} often rises to values even higher than those seen during the preictal period [9]. However, for sites which were less paticipating in the seizure transition, the STL_{max} values before and after the seizure are not significantly different. These temporal patterns for STL_{max} values are illustrated in Figure 2 and 3.

Most interestingly, Iasemidis observed that prior to each seizure, the signal from several electrode sites demonstrated a convergence in the values of STL_{max} [8]. We subsequently defined this as preictal dynamical entrainment [8, 9, 11, 10, 12]. Preictal entrainment appeared to evolve gradually over a period of approximately 30 minutes prior to each recorded seizure in a group of 5 patients [14]. During this evolution, there appeared to be a gradual increase in the number of entrained electrode sites as the time to seizure approached. We hypothesized that the seizure occurred as a result of dynamical entrainment of a critical number of cortical electrode sites. Interestingly, after each seizure, those electrodes that had become entrained in the preictal period were disentrained in the postictal period. Thus, the seizure appeared to serve as some sort of resetting mechanism.

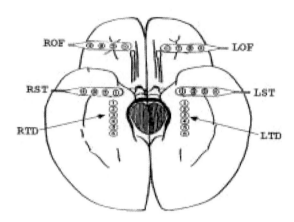

Figure 8.1. Schematic diagram of the depth and subdural electrode placement. This view from the interior aspect of the brain shows the approximate location of depth electrodes, oriented along the anterior-posterior plane in the hippocampi (RTD - right temporal depth, LTD - left temporal depth), and subdural electrodes located beneath the orbitofrontal and subtemporal cortical surfaces (ROF - right orbitofrontal, LOF - left orbitofrontal, RST- right subtemporal, LST- left subtemporal).

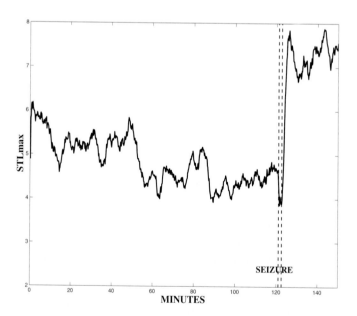

Figure 8.2. Smoothed STL_{max} profiles of a cortical site that participates in the seizure, over 150 minutes including a seizure. A gradual decline in STL_{max} preceded the seizure. During the seizure, STL_{max} falls to a minimum. Postictally, STL_{max} is reset to a higher value.

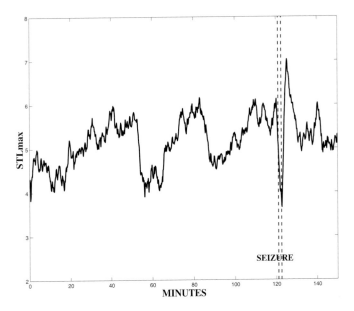

Figure 8.3. Smoothed STL_{max} profiles of a non-participating cortical site over 150 minutes including a seizure. In comparison to the site shown in figure 2, there is no preictal decline or postictal resetting of STL_{max} values.

Unfortunately, the preictal dynamical entrainment does not follow a simple and consistent pattern. Thus, it can not be studied with simple statistical techniques. For example, the specific sites that participated in the preictal entrainment varied from seizure to seizure. Our results indicate that electrode sites overlying the seizure onset zone (also known as the epileptogenic focus) are entraining other cortical sites, including sites contralateral as well as sites ipsilateral to the seizure onset zone. The rate of entrainment also varied from seizure to seizure. Thus, the spatiotemporal evolution of the preictal transition is highly variable. In order to study the complex spatiotemporal dynamics quantitatively, we needed to find a way to identify the critical cortical sites that participated in the preictal transition.

Further understanding of the process is enhanced by the combined use of novel dynamical measures in conjunction with sophisticated optimization techniques for selecting critical cortical sites. Using these techniques, we demonstrated that seizures are preceded by a gradual dynamical entrainment of widespread regions of both cerebral hemispheres [13, 14]. The entrainment is detectable by a convergence in the value of the short-term maximum Lyapunov exponent (STL_{max}) among critical cortical sites. This preictal (pre-seizure) transition evolves over minutes to hours. The preictal transition involves a subset of cortical sites, which vary from seizure to seizure [13]. In most instances one or more region of the epileptogenic focus interacts with other cortical sites. It appears as though the epileptogenic focus recruits other sites into an abnormal dynamical state. During the seizure itself, STL_{max} values of the participating sites drop to a minimum. Following each seizure, there is a resetting of spatiotemporal dynamics by the seizure. This resetting is characterized by a disentrainment of sites that were entrained during the preictal transition. This disentrainment is characterized by a divergence in the value of STL_{max} among electrode sites.

The combined application of nonlinear dynamical measures and quadratic optimization provides a powerful technique for studying state transitions in complex systems. We anticipated that these investigations will lead to a better understanding of the mechanisms involved in the occurrence of epileptic seizures.

By combining nonlinear dynamical methods with global optimization techniques, we have been able to demonstrate two important findings thus far: (1) almost all seizures are preceded by a preictal spatiotemporal dynamical transition detectable in the EEG signal [13, 14], (2) seizures appear to act as a resetting mechanism [15]. These observations have important implications about the mechanisms underlying the development and resolution of epileptic seizures. In addition, the ability to demonstrate these spatiotemporal dynamical transitions in an objective way provides the means for developing algorithms for automatically predicting seizures tens of minutes before they occur. If seizures can be predicted, it may be possible to develop novel approaches to seizure control, based on carefully timed interventions.

In this communication, we will provide a brief description of the dynamical measures we have employed in the study of human epilepsy. This will be followed by a brief discussion of the global optimization algorithm used to identify critical electrode sites. We will then provide evidence to support our hypothesis that seizures serve as a resetting mechanism, returning the brain to a more normal state.

The chapter is organized as follows: Section 2 provides a brief description of the nonlinear dynamical measure, STL_{max}, utilized in our research and its application to multichannel EEG time series. Section 3 introduces the zero-one global optimization technique and its application on the selection of critical channels to investigate the dynamical transition of the epileptic seizures. The hypotheses about the seizure being a resetting mechanism and the results of the statistical tests will be given in section 4. The final section gives the conclusion of this study.

2.   Nonlinear dynamical measures

The electroencephalogram (EEG) has been the most utilized signal to clinically detect brain function since its discovery by Richard Caton [2] and its first systematic investigation by Hans Berger [1, 4]. Unfortunately, based on simple assumptions about the system (e.g. linearity assumption), traditional signal processing technique has limited success of quantifying the characteristics of EEG. This limitation stems from the fact that the EEG is generated by a non-linear system, the brain. Therefore, it is necessary to develop signal processing techniques that will be capable of taking in consideration the nonlinear nature of the system under investigation. EEG characteristics such as α activity and seizures, instances of bursting behavior during light sleep, amplitude dependent frequency behavior (the smaller the amplitude the higher the EEG frequency) and existence of frequency harmonics (e.g. under photic driving conditions) are typical features of the EEG signal. These characteristics all belong to the long catalog of properties of typical nonlinear systems [16].

In addition to the above, the EEG time series, being the output of a multi-dimensional system, has statistical properties that depend upon both time and space [18]. Components of the brain (neurons) are heavily interacting and the EEG signals recorded from one area is inherently related to the activity in other areas. These components may be functionally connected at different time instants. Hence, the EEG should be considered a multivariate nonstationary time series. A well-established technique for visualizing the dynamical behavior of a multivariable, time-dependent system is to generate a state space portrait of the system. It is created by treating each time-dependent variable of the system as a component of a vector in a multidimensional space, usually called phase or state space of the system. Each vector in the state space represents an instantaneous state of the system. These time-dependent vectors are then plotted sequentially in the state space to represent the evolution of the state of the system over time. For many systems, this graphical representation creates an object confined over time to a sub-region of the state space. Such sub-regions of the state space are called "attractors". The geometrical properties of these attractors provide information about the steady state of the system.

One of the problems in the analysis of multidimensional systems is to determine which measurable observables of the system to analyze. Experimental constraints may limit the number of observables. However, when the variables of the system are dependent over time, which must be the case for any dynamical system to exist as a system, proper analysis of a single observable can provide information about the variables of the system that are related to this observation. Thus, one may be able to understand important features of a complex system through the study on a single observable over time.

In principle, through the method of delays described by Packard et al. [21] and Takens [25], sampling of a single observable over time can approximate the state of the system in a space spanned by the other dependent system variables. This method can be applied to reconstruct the state space from an EEG signal. Thus, a multidimensional state space can be created from a single-electrode

EEG recording. In such an embedding, each state is represented in the state space by a vector $x(t)$ whose components are the delayed versions of the original single-channel EEG time series $u(t)$, that is:

$$x(t) = [u(t), u(t - \tau), ..., u(t - (p - 1) \times \tau)],$$

where $x(t)$ is a vector in the state space at time t, τ is the time delay between successive components of $x(t)$, and p is the embedding dimension of the reconstructed state space.

According to Takens, the embedding dimension p should be at least equal to $(2D + 1)$ in order to correctly embed an attractor in the state space, where D is the dimension of the attractor. The measure most often used to estimate D is the state space correlation dimension ν. Methods for calculating ν from experimental data have been described [19, 17] and were employed in our work to approximate D of the epileptic attractor. Through the EEG data we have analyzed, ν is found to be between 2 and 3 during the ictal period of an epileptic seizure. Therefore, in order to capture characteristics of the epileptic attractor, we have used an embedding dimension p of 7 for the reconstruction of the state space.

An attractor is chaotic if, on the average, orbits originating from similar initial conditions (nearby points in the state space) diverge exponentially fast. If these orbits belong to an attractor of finite size, they will fold back into it as time evolves. The result of these two processes may be a stable, topologically layered, attractor [5]. When the expansion process, on average, overcomes the folding process in some eigendirections of the attractor, the attractor is called chaotic.

Since the brain is a nonstationary system, algorithms used to estimate measures of the brain dynamics should be capable of automatically identifying and appropriately weighing existing transients in the data. The method we developed for estimation of L_{max} for nonstationary data, called STL (Short Time Lyapunov), considers possible nonstationarities in the EEG. This method is explained in detail in Iasemidis et al. [9, 7]. We apply the STL algorithm to EEG tracings from electrodes in multiple brain sites (see Figure 1), to create a set of STL_{max} time series. This time series contains local (in time and in space) information about the brain as a dynamical system. Figures 2 and 3 show the STL_{max} profiles of two electrodes over 150 minutes (120 minutes before and 30 minutes after a seizure). We can see that a significant drop occurs during the seizure period for both electrodes. When comparing the STL_{max} values before and after the seizure, one electrode has significantly larger values of STL_{max} after the seizure, the other one remains at the same level as before the seizure. Since STL_{max} measures the chaocity of the system, these observations show that the system of the brain is less chaotic during the seizure (itcal) period, and for some area of brain, it returns to the more chaotic state after the seizure (postictal) compared with the state before the seizure (preictal). It will be shown that it is at this level of spatiotemporal analysis (i.e., multi-electrode temporal analysis of the dynamics) that reliable detection of the transition to epileptic seizures, long before they actually occur, is derived.

3. Zero-one global optimization

For many years the Ising model [3, 24] has been a powerful tool in studying phase transitions in statistical physics. Such an Ising model can be described by a graph $G(V, E)$ having n vertices $\{v_1, \ldots, v_n\}$ and each edge $(i, j) \in E$ having a weight (interaction energy) J_{ij}. Each vertex v_i has a magnetic spin variable $\sigma_i \in \{-1, +1\}$ associated with it. A spin configuration of minimum energy is obtained by minimizing the Hamiltonian $H(\sigma) = -\sum_{1 \le i < j \le n} J_{ij} \sigma_i \sigma_j$ over all $\sigma \in \{-1, +1\}^n$. This problem is equivalent to the combinatorial problem of quadratic bivalent programming [6].

Motivated by the application of the Ising model to phase transitions we have used quadratic bivalent (zero-one) programming for the optimal selection of brain sites at periods prior to epileptic seizures [13, 14]. The objective function to be minimized was the distance of measures of chaos (STL_{max}) between recording brain sites. These measures were estimated as described before. The sites selected by the optimization method have provided two important insights. First, sites participating in the preictal transition could thus be identified. Although these sites differ from seizure to seizure, the sites that are most frequently selected are located in the epileptogenic zone. Second, convergence of dynamical measures of the selected sites over time could detect the epileptic transition and predict the upcoming epileptic seizure well before its onset, a phenomenon we have called dynamical entrainment [13, 14].

More specifically, we considered the integer 0-1 problem:

$$\min (x'\boldsymbol{T}x) \text{ with } x \in \{0, 1\}^n \text{ subject to the constraint } \sum_{i=1}^{n} x_i = k, \quad (1)$$

where n is the total number of electrode sites and k the number of sites to be selected. The elements of the matrix $\boldsymbol{T} = (\boldsymbol{T}_{ij})$ are statistical measures of the distances of brain sites i and j with respect to the estimated mean and standard deviation of their STL_{max} values within 10 minute windows W. The statistical measures of distance we have used in this analysis are the T-indices from the well known t-test in statistics and are described in the next section. If the constraint in (1) is included in the objective function $f(x) = x'\boldsymbol{T}x$ by introducing the penalty

$$\mu = \sum_{j=1}^{n} \sum_{i=1}^{n} \boldsymbol{T}_{ij} + 1,$$

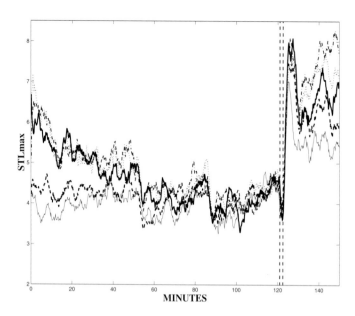

Figure 8.4.   Smoothed STL_{max} profiles of 5 optimal electrode sites over 150 minutes including a seizure. The preictal period shows gradual convergence of the STL_{max} values calculated
for these critical electrode sites. During the seizure, STL_{max} values are completely entrained.
Postictally, the values are disentrained indicating resetting which reverses the preictal entrainment.

the optimization problem becomes equivalent to an unconstrained global optimization problem:

$$\min \left[x' \mathbf{T} x + \mu \left(\sum_{i=1}^{n} x_i - k \right)^2 \right], \text{ where } x \in \{0, 1\}^n \qquad (2)$$

The electrode site i is selected if $x_i = 1$ in the solution of (2). Led by a variety of
empirical correlations and numerical experiments we have chosen the value of
$k = 5$ electrode sites to be selected (i.e. the five most entrained electrode sites)
as a balance between sensitivity and specificity. Higher values of k decrease
specificity whereas lower values of k decrease the sensitivity of the algorithm.
    Figures 4 and 5 show the STL_{max} profiles of two sets of five channels over
the same time interval as in figures 2 and 3. The set of five channels in figure 4
were selected by applying the optimization technique in a 10 minutes window
before the seizure, and the other set of five channels in figure 5 were not selected
optimally. Clearly, the STL_{max} values of the optimal set of channels in figure
4 gradually converge (entrain) before the seizure and then diverge (disentrain)
right after the seizure. But the other set of channels are close together before and
after the seizure. Other non-optimal sets of channels could be apart before and

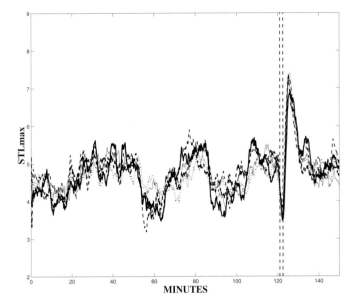

Figure 8.5. Smoothed STL_{max} profiles of 5 non-optimal electrode sites over 150 minutes including a seizure. Postictal resetting is not observed for these sites.

after the seizure, or apart before and close together after the seizure. This observation indicates that by following the change of the STL_{max} values among the group of critical channels, it is possible to observe the dynamical transitions (entrainment before the seizure and disentrainment after the seizure) of a epileptic seizure and even to predict it long before it happens. The introduced optimization technique appears to be a possible solution of how to identify those critical channels.

4. Statistical testing of the resetting hypotheses

Based on qualitative analysis of a small sample of seizure recordings [22, 23], we postulated that a seizure occurs after the STL_{max} values of a group of critical electrode sites become dynamically entrained and that this entrainment among those critical sites is dynamically reset by the seizure. In this section, we present results from analysis of 22 seizures in 1 patient that support the following hypotheses:

Hypothesis I Resetting is detectable in a specific subset of electrode sites and at seizure points.

Hypothesis II Resetting occurs more frequently at seizure points than other time points in interictal periods.

The patient was a 41 year old right handed female who had a history of an intractable sizure disorder since the age of eight years following measles infection. She had been considered medically intractable since the age of 11. Macroelectrodes had been surgically implanted in the hippocampus, temporal and frontal lobe cortex, bilaterally (Figure 1). Approximate 9 days (217.4 hours) of intracranial continuous EEG multi-channel recordings (30 common reference channels) was analyzed to test the two hypotheses we postulated. Twenty-two seizures of mesial temporal onset were recorded for this patient during the period of recordings.

Based on the STL_{max} of all n channels ($n = 30$), the statistical distance T_{ij} (T-index) of any pair (i, j) of channels within a $\tau = 10$ minute window W was calculated over time. Specifically, using the pairwise statistical approach (t-distribution), the T- index at time t between electrode sites i and j was defined as:

$$T_{i,j}(t) = |E\{STL_{max_i}(t) - STL_{max_j}(t)\}| \cdot \sqrt{N}/\sigma_{i,j}(t),$$

where $E\{\cdot\}$ denotes the average of all differences $STL_{max_i}(t) - STL_{max_j}(t)$ within a moving window $W_t(\lambda)$ defined as:

$$W_t(\lambda) = \begin{cases} 1 & \text{if } \lambda \in [t - \tau, t] \\ 0 & \text{if } \lambda \notin [t - \tau, t] \end{cases}$$

and N is the number of STL_{max} values estimated within the moving window W_t. Then, $\sigma_{i,j}(t)$ is the sample standard deviation of the STL_{max} differences between electrode sites i and j within the moving window $W_t(\lambda)$. The thus defined T-index follows a t-distribution with $N - 1$ degrees of freedom.

In the estimation of the $T_{i,j}(t)$ indices in our data we used $N = 60$ (i.e. averages of 60 differences of STL_{max} exponents per moving window for a pair of electrode sites i and j). Since each exponent in the STL_{max} profiles is derived from a 10.24 second EEG data segment, the length of the window used corresponds to approximately 10 minutes.

By inputting the thus formed T-index $(n \times n)$ matrix T to the optimization program that solves equation (2) above, the 5 most entrained sites at time t (i.e. within [t-10 min, t]) are selected. These sites form a set of 5 electrode sites for which the objective function attains its minimum over all possible sets of 5 sites. After site selection, the entrainment and disentrainment periods of the selected sites, prior to and after time t respectively, were estimated as follows. First, the average of the T-indices of all pairs of the selected sites over time (i.e., moving $W_t(\lambda)$ forward and backward in time with respect to t) is estimated. Thus, spitially averaged T-index curve over time and critical sites is generated (see Figures 6 and 7). We then define the entrainment period (T_e) as the period before t during which the T-index curve of the selected sites remains continuously below 2.662, which is the 1% statistical significance level from the t-distribution. The disentrainment period (T_d) is defined as the period after

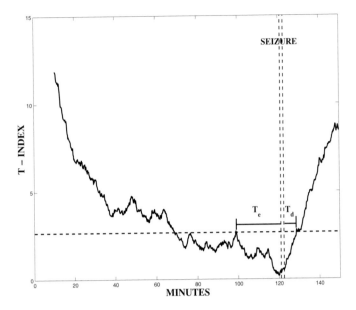

Figure 8.6. T-indes curve over time, before and after the seizure depicted in figure 3, based on the average of the STL_{max}'s of the electrode sites optimally selected in a 10 minute window before that seizure.

t during which the T-index curve of the selected sites remains continuously below 2.662.

In figure 6, the averaged T-index curve was generated for the sites in figure 4, which were optimally selected by the optimization algorithm as described in the previous section. We can see that the T-index curve slowly drops below 2.662 (dashed line) about 30 minutes before the seizure and remains entrained and remains there during the seizure. After the seizure, the T-index values become larger than the critical value 2.662 very fast(< 10 minutes) and stay disentrained thereafter. This is a typical epileptic seizure transition observed by the T-index curve of the STL_{max} values in this patient. The transition includes the entrainment of the STL_{max} values among the critical sites before a seizure and the disentrainment after the same seizure. For the average T-index curve in figure 7 (generated for the non-optimally selected sites in figure 5), this transition is not observed. This comparison shows the importance of the optimization technique.

We now define a resetting occurring at time t if $T_d(t) < T_e(t)$, that is when the disentrainment period is shorter than the entrainment period for the critical sites selected at time t. If p is the probability of a resetting at a time point t, and if we denote by x_m the random variable of the number of resettings observed at m time points $t_1, t_2, \ldots, t_m$, then x_m follows the binomial(m, p) distribution,

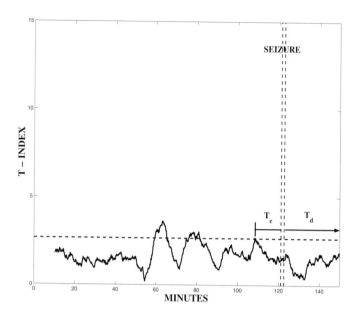

Figure 8.7.   T-indes curve over time, before and after the seizure depicted in figure 4, based on the average of the STL_{max}'s of the electrode sites in figure 5.

and p can be easily estimated by the maximum likelihood estimator as

$$p^* = (\text{number of points where } T_d < T_e)/m = x_m/m$$

For this patient, the average and standard deviation of T_e and T_d periods across seizures as well as the p^* values of resetting at seizure point using the optimally selected set of sites are estimated ($T_e = 35.72 \pm 2.56; T_d = 10.88 \pm 0.46; p^* = 1.00$). According to these results, all seizures (100%) reset the preictal entrainment. As a byproduct of this test, the duration of the preictal transition (existence of dynamical entrainment prior to an epileptic seizure) is derived. In particular, the entrainment period (T_e) was approximately 35 minutes and overall significantly larger than the corresponding disentrainment period (T_d) with p-value < 0.0001.

To show the significance of optimization (selection of sites) in the detection of seizure resetting, we compared the results on resetting at seizure points using the optimal set of sites versus uniformly random selected sites. Thus, we first define $p_{s,opt}$ = probability of resetting by a seizure using the optimal combination of sites, and $p_{s,n-opt}$ = probability of resetting by a seizure using other sets of sites. Then we test null hypothesis $H_0 : p_{s,opt} = p_{s,n-opt}$ versus alternative hypothesis $H_a : p_{s,opt} > p_{s,n-opt}$ by comparing the observed $p^*_{s,opt}$ with the distribution of the $p^*_{s,n-opt}$ values for many non-optimal groups of electrodes. This distribution was estimated by 25,000 simulations, with each

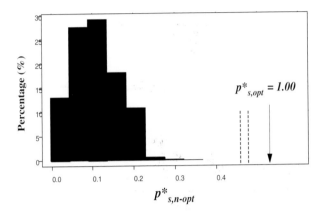

Figure 8.8. Probability distribution of resetting at seizure points using non-optimal group of electrode sites.

simulation randomly selecting the same number ($k = 5$) of sites at each of the 22 seizures. Increasing the number of simulations did not change the statistical significance of the final results. Our results (see Figure 8) show that, at the 95% confidence level, we can reject H_0 (p-value < 0.00004) in this patient. Hence the probability of resetting by a seizure using the optimal combinations of sites is, statistically, significantly greater than that using non-optimal groups of sites. Therefore, our first hypothesis (*resetting should be detectable at seizure points in a specific subset of electrode sites*) is confirmed in this patient.

The second hypothesis (*resetting should occur most frequently when seizures are observed in comparison to interictal periods*) was also confirmed.

Having shown above that optimal selection of brain sites increases the sensitivity of detection of resetting, we test hypothesis II using the optimally selected sites at any time point t. First we define $p_{s,opt}$ = probability of resetting for the optimal combination of sites at the seizure points, and $p_{n-s,opt}$ = probability of resetting for the optimal combination of sites at the non-seizure points. Then we test null hypothesis $H_0 : p_{s,opt} = p_{n-s,opt}$ versus alternative hypothesis $H_a : p_{s,opt} > p_{n-s,opt}$ by comparing the observed $p^*_{s,opt}$ with the distribution of the $p^*_{n-s,opt}$ values for time points in non-seizure periods. This distribution is now estimated by 25,000 simulations, with each simulation randomly selecting 22 time points in non-seizure periods within the whole (approximate 10 days in duration), continuous EEG record. Then, at the 95% confidence level, we

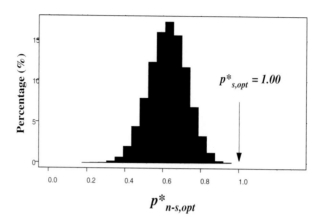

Figure 8.9.   Probability distribution of resetting at the non-seizure points using optimal group of electrode sites.

can reject H_0 if the p-value of the test is less than 0.05. For this patient, the probability of resetting at the 22 seizure points (100%) was significantly greater than that at 22 randomly selected points in interictal periods, where resetting is observed with an average probability of 63% (see Figures 9).

5.   Conclusion

Investigation of brain dynamics through analysis of multi-channel, multi-day EEG recordings in one patient with multiple epileptic seizures, demonstrated that: 1) critical cortical sites become dynamically entrained on the order of tens of minutes prior to a seizure and are dynamically disentrained significantly faster following epileptic seizures. The hypotheses that resetting occurs at seizure points and involves a critical subset of cortical sites (Hypothesis I) and that resetting is specific to seizure occurrence (Hypothesis II) were confirmed at the 95% confidence level. 2) The criterion for resetting at a time point t was that the entrainment period before t is larger than the corresponding disentrainment period after t. The variability in participating sites, as well as in entrainment and disentrainment periods at a seizure point, makes it difficult to detect the phenomenon. These results were made possible by employment of a novel methodology combining optimization, techniques from nonlinear dynamics, and statistics.

References

[1] H. Berger. Uber das elektroenkephalogramm des menchen. *Arch. Psychiatr. Nervenkr.*, 87:527–570, 1929.

[2] R. Caton. The electric currents of the brain. *BMJ*, 2:278, 1875.

[3] C. Domb. In C. Domb and M. S. Green, editors, *Phase Transitions and Critical Phenomena*. Academic Press, New York, 1974.

[4] P. Gloor. *Hans Berger on the electroencephalogram of man.* Elsevier, Amsterdam, 1969.

[5] A. V. Holden. *Chaos-nonlinear science: theory and applications.* Manchester University Press, Manchester, 1986.

[6] H. Horst, P. M. Pardalos , and V. Thoai. *Introduction to global optimization, second edition, Series on Nonconvex Optimization and its Applications, 3.* Kluwer Academic Publishers, Dordrecht, 2000.

[7] L. D. Iasemidis, J. C. Sackellares, H. P. Zaveri, and W. J. Williams. Phase space topography of the electrocorticogram and the Lyapunov exponent in partial seizures. *Brain Topogr.*, 2:187–201, 1990.

[8] L. D. Iasemidis. *On the dynamics of the human brain in temporal lobe epilepsy.* Ph.D. thesis, University of Michigan, Ann Arbor, 1991.

[9] L. D. Iasemidis and J. C. Sackellares. The temporal evolution of the largest Lyapunov exponent on the human epileptic cortex. In D. W. Duke and W. S. Pritchard, editors, *Measuring chaos in the human brain.* World Scientific, Singapore, 1991.

[10] L. D. Iasemidis, J. C. Principe, and J. C. Sackellares. Spatiotemporal dynamics of human epileptic seizures. In R. G. Harrison, W. Lu, W. Ditto, L. Pecora, M. Spano, and S. Vohra, editors, *3rd Experimental Chaos Conference.* World Scientific, Singapore, 1996.

[11] L. D. Iasemidis and J. C. Sackellares. Chaos theory and epilepsy. *The Neuroscientist*, 2:118–126, 1996.

[12] L. D. Iasemidis, J. C. Principe, J. M. Czaplewski, R. L. Gilman, S. N. Roper, and J. C. Sackellares. Spatiotemporal transition to epileptic seizures: A nonlinear dynamical analysis of scalp and intracranial EEG recordings. In F. Lopes da Silva, J. C. Principe, and L. B. Almeida, editors, *Spatiotemporal Models in Biological and Artifical Systems*. IOS Press, Amsterdam, 1997.

[13] L. D. Iasemidis, D. S. Shiau, J. C. Sackellares, and P. M. Pardalos. Transition to epileptic seizures: Optimization. In D. Z. Du, P. M. Pardalos and J. Wang, editors, *DIMACS series in Discrete Mathematics and Theoretical Computer Science, vol. 55*. American Mathematical Society, 1999.

[14] L. D. Iasemidis, P. M. Pardalos, J. C. Sackellares, and D. S. Shiau. Quadratic binary programming and dynamical system approach to determine the predictability of epileptic seizures. *Journal of Combinatorial Optimization*. 5:9-26, 2000.

[15] L. D. Iasemidis, P. M. Pardalos, J. C. Sackellares, and D. S. Shiau. Global optimization and nonlinear dynamics to investigate complex dynamical transitions: Application to human epilepsy. *IEEE Transactions on Biomedical Engineering*. in press.

[16] B. H. Jansen. Is it and so what? A critical review of EEG-chaos. In D. W. Duke and W. S. Pritchard, editors, *Measuring chaos in the human brain*. World Scientific, Singapore, 1991.

[17] E. J. Kostelich. Problems in estimating dynamics from data. *Physica D*, 58:138–152, 1992.

[18] F. Lopes da Silva. EEG analysis: theory and practice; Computer-assisted EEG diagnosis: Pattern recognition techniques. In E. Niedermeyer and F. Lopes da Silva, editors, *Electroencephalography: Basic principles, clinical applications and related field*. Urban and Schwarzenberg, Baltimore, 1987.

[19] G. Mayer-Kress. *Dimension and entropies in chaotic systems*. Springer-Verlag, Berlin, 1986.

[20] M. Palus, V. Albrecht, and I. Dvorak. Information theoretic test for nonlinearity in time series. *Phys. Lett. A*, 175:203–209, 1993.

[21] N. H. Packard, J. P. Crutchfield, J. D. Farmer, and R. S. Shaw. Geometry from time series. *Phys. Rev. Lett.*, 45:712–716, 1980.

[22] J. C. Sackellares, L. D. Iasemidis, R. L. Gilman, and S. N. Roper. Epileptic seizures as neural resetting mechanisms. *Epilepsia*, 38 S3: 189, 1997.

[23] D. S. Shiau, Q. Luo, R. L. Gilmore, S. N. Roper, P. M. Pardalos, J. C. Sackellares, and L. D. Iasemidis. Epileptic seizures resetting revisited. *Epilepsia*, 41 S7:208, 2000.

[24] D. L. Stein. In D. L. Stein, editor, *Lecture Notes in the Sciences of Complexity, SFI Studies in the Science of Complexity*. Addison-Wesley Publishing Company, 1989.

[25] F. Takens. Detecting strange attractors in turbulence. In D. A. Rand and L. S. Young, editors, *Dynamical systems and turbulence, Lecture notes in mathematics*. Springer-Verlag, Heidelburg, 1981.

Chapter 9

FUNCTIONAL MAGNETIC RESONANCE IMAGING DATA ANALYSIS WITH INFORMATION-THEORETIC APPROACHES

Qun Zhao
NIMH Center for Study of Emotion and Attention*, *University of Florida*
zhao@cnel.ufl.edu

Jose Principe
Department of Electrical and Computer Engineering, University of Florida
principe@cnel.ufl.edu

Jeffery Fitzsimmons
Department of Radiology, University of Florida

Margaret Bradley
NIMH Center for Study of Emotion and Attention, University of Florida

Peter Lang
NIMH Center for Study of Emotion and Attention,University of Florida

Abstract   An information-theoretic approach is presented for functional Magnetic Resonance Imaging (fMRI) analysis to detect neural activations. Two divergence measures are employed to estimate the difference between two data distributions, obtained by segmenting the voxel time series based on the experimental protocol time-line. In order to validate the new technique activation signals were acquired from a specially constructed dynamic fMRI phantom placed in

* Supported by National Institute of Mental Health grants MH52384, MH37757 and MH43975, and National Science Foundation grant ECS-9900394

P.M. Pardalos and J. Principe (eds.), Biocomputing, 159-173.
© 2002 *Kluwer Academic Publishers. Printed in the Netherlands.*

the scanner, instead of computer simulated signals. The data analysis shows that the divergence measure is more robust in quantifying the difference between two distributions than the methods that utilize only the first and second order statistics. More importantly, the divergence measure is a method for calculating the brain activation map that makes no assumptions about the data distribution (e.g., Gaussian distribution). Results are also shown based on data of visual task study of subjects.

Keywords: fMRI, parametric, nonparametric, information-theoretic analysis

Introduction

Many different methods from both signal processing and statistics have been applied in recent years to fMRI analysis. Generally, these techniques can be divided into two groups, parametric (such as statistical t-test, cross-correlation, and general linear model) and non-parametric approaches (e.g.,Kolmogorov-Smirnov test) (Friston et.al., 1995; Lange, 1999; Worsley & Friston, 1995; Worsley, et.al.,1996).

The advantage of the parametric approaches is its simplicity. However, for example, the t-test utilizes just the first and second order statistics, i.e., the mean and pooled variance, based on the assumption of Gaussian distribution for the data set. When the statistics of the generative mechanism and/or the size of data set or degree of freedom (DOF) are limited, this Gaussian assumption may not be justified (as shown in later study), or in other words the t-test is an inaccurate approximation. Non-parametric methods make no assumptions about the possible parametric families of distributions generating the fMRI data and are thus less dependent on a specific statistical model.

In this chapter, we propose a new non-parametric method, mutual information analysis, based on information theory (Kullback, 1968; Shannon, 1948). Information-theoretic methods are becoming more and more popular in recent years, which has been used in fMRI analysis (Tsai, et.al.,1999), combining multiple source imaging (Baillet & Garnero, 1999), and multi-modality image registration (Maes, et.al.,1997). In the paper by Tsai et.al., their method is based on a formulation of the mutual information between two waveforms, the fMRI temporal response of a voxel and the experimental protocol time-line. Scores based on mutual information were generated for all voxels and then used to compute the activation map of the experiment. It was shown that mutual information is robust in quantifying the relationship between any two waveforms, but no numerical evidence is presented on the power of the test. In general, the advantage of the information-theoretic approach is that it makes no assumptions about the linearity of input/output relationships (e.g., general linear model) or the normality of the data distribution.

In this chapter we first provide a short introduction to information-theoretic methods. Two mutual information based divergence measures are presented. The data used were obtained from a specially constructed phantom in which functional effects are modeled based on known physical parameters, and data from human subjects performing a perceptual task.

1. Information-theoretic approaches

In information theory (Shannon, 1948; Kullback,1968; Kapur & Kesavan,1992) entropy and mutual information are two important concepts. Simply speaking, entropy is a measure of the average uncertainty of a random variable, while mutual information is a measure of the information that one random variable or distribution conveys about another. In fMRI, entropy can be used to describe one data distribution (or data under one condition), while mutual information is used to find the differences between two data distributions (e.g, data acquired under two conditions). The reader is referred to (Kullback,1968; Kapur & Kesavan,1992) for a complete treatment. The earliest definition of the (differential) entropy of a continuous random variable x was proposed by Shannon (Shannon, 1948)

$$H_s(x) = -\int f(x)log(f(x))dx \qquad (1)$$

where $f(x)$ is the probability density function (PDF) of x. The mutual information or cross-entropy between two random variables x and y can be written as

$$\begin{aligned} I(x,y) &= H(x) - H(x|y) = H(y) - H(y|x) \qquad (2) \\ &= H(x) + H(y) - H(x,y) \end{aligned}$$

where $H(x|y)$ and $H(y|x)$ are the conditional entropies, and $H(x,y)$ is the joint entropy. The relationship between them is $H(x,y) = H(x|y) + H(y) = H(y|x) + H(x)$.

It is noted that Shannon's entropy is not easily estimated because of the integral of the product of $f(x)$ with its logarithmic form. Shannon entropy (1) can only be utilized in an approximate sense to estimate the mutual information in (2). Other forms of entropy, which are easier to estimate, include Renyi entropy H_R and Havrada-Charvart H_{hc} entropy, and are defined as (Kapur & Kesavan, 1992)

$$H_R(x) = \frac{1}{1-\alpha} log \left(\int f^\alpha(x) dx \right) \qquad (3)$$

and

$$H_{hc}(x) = \frac{1}{1-\alpha} \left(\int f^\alpha(x) dx - 1 \right) \qquad (4)$$

where α is a parameter that controls the order of entropy. We prefer Renyi's entropy because of its computational simplicity in measuring the divergence between two probability distributions. A general definition of mutual information utilizes the Kullback-Leibler (KL) divergence (or relative entropy), which measures the divergence between two probability distributions $f_1(x)$ and $f_2(x)$, as

$$K(f_1(x) \| f_2(x)) = \int f_1(x) log \frac{f_1(x)}{f_2(x)} dx \qquad (5)$$

The Kullback-Leibler divergence is non-negative, and is zero if and only if the two distributions, f_1 and f_2, are the same. In this sense it can be regarded as a "distance" measure between two distributions. However, it is not symmetric, i.e., $K(f_1(x) \| f_2(x)) \neq K(f_2(x) \| f_1(x))$. Therefore, it is more properly called a divergence or "directed distance".

Although the KL divergence has many useful properties, it requires that probability distributions satisfy the condition of absolute continuity (Kullback, 1968). Also, it is difficult to estimate from samples due to the presence of the integral and the logarithmic function in Eqn.(5).

2.   Two alternative divergence measures

Here we propose to utilize alternative distance measures to replace the KL divergence between two PDFs. The objective is to investigate the difference between two PDFs to find if they satisfy either of the following hypotheses

$$H_0 : f_1 = f_2$$
$$H_1 : f_1 \neq f_2$$

Then a natural statistic of measuring the difference is the integrated square error (ISE) between two distributions Anderson (1994),

$$Q = \int (f_1(x) - f_2(x))^2 dx \qquad (6)$$

In the real world, usually we only have discrete data points, therefore we have to estimate the distance measure Q based on these discrete data. Given two sample sets $\{X_{j1}, ..., X_{jn_j}\}, j = 1, 2$, first we start with estimation of their PDFs $f_j(x), j = 1, 2$ by the Parzen window method (Parzen, 1968),

$$\hat{f}_j(x) = (n_j d_j^p)^{-1} \sum_{i=1}^{n_j} K\{\frac{x - X_{ji}}{d_j}\}, j = 1, 2 \tag{7}$$

where d_j is the kernel size or bandwidth and K is a spherically p-variate density function, e.g., the Gaussian function. Then Q can be estimated by a statistic T_{ISE} as

$$T_{ISE} = \int (\hat{f}_1(x) - \hat{f}_2(x))^2 dx \tag{8}$$

In the following we will examine T_{ISE} for testing the two hypothesis H_0 and H_1. In order to assess the power of a test based on T_{ISE} we can ascertain its performance against a local alternative hypothesis. Let $f_1 = f$ denote a fixed density, and let g be a function such that $f_2 = f + \delta g$ is also a density for all sufficiently small $|\delta|$, then we can set the significance of the statistic. Let t_d denote the α-level threshold of the distribution of T_{ISE} under the null hypothesis, i.e., $\delta = 0$, the test statistic consists of rejecting H_0 if $T_{ISE} > t_d$, i.e.,

$$P_{H_0}(T_{ISE} > t_d) = \alpha$$

Anderson proved that if d, the kernel size, is chosen to converge to zero as $n_1, n_2 \to \infty$, then the minimum distance at which the test statistic can discriminate between f_1 and f_2 is $\delta = (n_1 + n_2)^{-1/2} h^{-p/2}$ (Anderson, 1994). Furthermore, Diks proposed T_{ISE} as an unbiased estimator of smoothed Q using Gaussian kernels (Diks, 1996). This same methodology was proposed for machine learning application by our group(Fisher & Principe, 1997) and later shown to be equivalent to an approximation to K-L divergence using Renyi's quadratic entropy (Principe el al.,2000).

Besides the ISE divergence measure, another alternative measurement of divergence, based on the Cauchy-Schwartz inequality,

$$(\int f_1(x) f_2(x) dx)^2 \leq (\int f_1^2(x) dx)(\int f_2^2(x) dx)$$

was proposed in (Principe el al., 2000), and leads to the following statistic

$$T_{SI} = 1 - \frac{(\int f_1(x) f_2(x) dx)^2}{(\int f_1^2(x) dx)(\int f_2^2(x) dx)} \tag{9}$$

Here, it should be noted that the defined divergence, T_{SI} is within the range of $[0, 1]$. T_{ISE} and T_{SI} are both easy to estimate using the Parzen window PDF estimation and the quadratic form of Renyi's entropy which has been called the information potential (Principe et al.,2000) and reads,

$$H_{R2}(x) \;\; = \;\; -logV(x) \tag{10}$$

$$V(x) \;\; = \;\; \frac{1}{N^2}\sum_{i=1}^{N}\sum_{j=1}^{N}K\{\frac{X_i - X_j}{\sigma}\} \tag{11}$$

where $H_{R2}(x)$ is the Renyi's quadratic entropy and $V(x)$ is the information potential.

3.   fMRI neural activation study

3.1.   Dynamic fMRI phantom simulation

Systematic comparison of the analysis methods of fMRI in the literature can not be conclusive if assessment is based only on the highly variable output of the human brain. Calibrated, repeatable fMRI signals are needed for a reliable test. In effort to deal with this problem researchers have tried to numerically simulate the activation signals (Sorenson & Wang,1996; Constable, et.al, 1998). In (Sorenson & Wang, 1996), each time course data was represented by a zero-baseline boxcar function. Gaussian noise was added to simulate physiologic noise, after which the data were convolved with a Poisson function to simulate hemodynamics smoothing and delay. Also, structured noise in the form of slopes to simulate signal drift and sinusoidal oscillations to simulate respiratory motion were added. These simulated fMRI responses were finally added to null data sets acquired on normal subjects, denoted as "human data". In (Constable, et.al, 1998) Gaussian-shaped activation signals were added with Gaussian distribution of white noise. As indicated in (Le & Hu, 1997), these simulations, while providing a well modeled "true" activation signal, do not account for variations in the activation levels of different pixels and are unable to depict accurately the temporal response of the activation associated signal change. To overcome these limitations, a dynamic fMRI phantom was developed that generates simulated functional changes based on known, controllable signal input. When placed in the scanner the phantom produces calibrated, repeatable fMRI activities. This unique device permitted an objective, systematic comparison of mutual information based divergence and t-statistic methodologies.

A dynamic fMRI phantom was designed to simulate the activation signal in the human brain, which produces a signal intensity change within the physiological range. The phantom, which is a bottle filled with water, has a tuned radio-frequency (RF) coil attached to its surface. The basic principle is that the current passing through the RF coil induces the change of local magnetic field, which results in the changes of signal intensity when the resistance of the coil is changed. The signal intensity change is defined as,

$$R_{IC} = \frac{\bar{x}_{on} - \bar{x}_{off}}{\bar{x}_{off}} \tag{12}$$

where $\bar{x}_{on}$ and $\bar{x}_{off}$ are the mean values of signals acquired when the RF coil was turned on and off, respectively. The two pictures in Figures 9.1 and 9.2 give examples of a voxel time series with two different R_{IC}, one producing a change of 3.11 percent and another higher 10.67 percent.

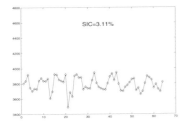

Figure 9.1. Voxel time series with low signal intensity changes.

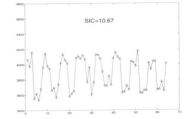

Figure 9.2. Voxel time series with high signal intensity changes.

It is worth mentioning that the voxel time series show qualitatively similar characteristics of the activation in real fMRI time series. The only difference is that we know exactly the time series signal because we control the switching of the current in the phantom.

Sensitivity and *Specificity* are two important factors in evaluating performance of a detection methodology. An appropriate way of assessing the two factors is by means of the receiver operating characteristic (ROC) analysis, which is based on the Neyman-Pearson criterion of statistical detection theory (Helstrom, 1968). The Neyman-Pearson criterion gives an appropriate estimate of performance in situations when an estimate of the prior probabilities of the hypotheses is not available, and the costs or risks attending the various decisions are not accurately known. This is exactly the case in fMRI analysis, where previous studies had applied the ROC analysis (Sorenson & Wang, 1996; Le & Hu, 1997; Constable, et.al, 1998). The ROC curve is generated by plotting the true positive fraction (TPF) against the false positive fraction (FPF). Based on the dynamic fMRI phantom data, we compared the performance of the nonparametric divergences (T_{ISE}, T_{SI}) and the parametric t-statistics (independent and dependent).

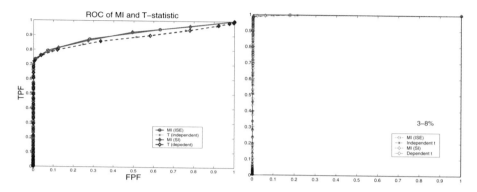

Figure 9.3.   ROC curve of nonparametric divergences (solid) and parametric t-statistics (dash). Left: R_{IC} is below 3 percent. Right: R_{IC} is from 3 to 8 percent.

ROC curves were computed respectively at the lower ratio of intensity change (R_{IC} is below 3 percent) and the higher range (3 to 8 percent). The two solid lines represent the nonparametric divergences (T_{ISE} and T_{SI}) and the two dash lines describe the parametric t statistics (independent and dependent). The result for low intensity change is shown in Figure 9.3 (left) and for the broader range in Figure 9.3 (right). We observe that for the low intensity change, the performance of the nonparametric divergences are better than that of the parametric t-statistics, but the individual types do not differ within groups. In Figure 9.3 (right) we see that the ROCs for all the methods are comparable despite high intensity range. These experiments show that the nonparametric divergences will produce more reliable estimate of activation, although improvements are still needed in the low intensity range.

3.2.   Study on human subject data

In this experiment, activation in the visual cortex was studied. Stimuli were several sets of pleasant, neutral, and unpleasant color pictures selected from the International Affective Picture System (Lang, Bradley, & Cuthbert, 1997) and were presented at a 3 Hz flashing rate. There were six blocks in the experiment, each consisting of eight cycles of a 12-s inter-trial interval (*off*) and a 12-s stimulus presentation period (*on*).

The functional images were acquired using a gradient echo EPI technique on a 3.0T GE Signa scanner. The functional scans were 6mm thick with a 0.5mm inter-slice gap, 128 X 128 pixels, 20mm field of view, and a flip angle of 90 degrees. Six coronal images (approximately 3 s/image acquisition) were collected during each of eight 12-s picture presentation periods and eight 12-s inter-picture periods in each block of trials, resulting in a set of 64 images at each coronal location per block (32 during stimulus presentation, 32 during inter-trial intervals).

Functional activity in each of six coronal images was determined as follows. Using the set of 64 images (32 *on*, 32 *off*) collected in the experiment, the parametric t-statistic and the nonparametric divergence T_{ISE} were used and their results were compared. To take into account the phase lag in blood flow, a lag of two images (approximately 6s) was considered (Lang, et.al, 1998).

Figure 9.4 gives an illustration of two voxel time series from our fMRI study and the estimate of the corresponding signal intensity PDF using a Parzen window estimation (7). We conclude that the two distributions are far from the Gaussian distribution and are not independent either. Instead, they show multimode distributions and are correlated to a certain degree.

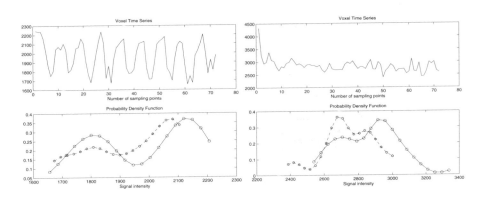

Figure 9.4. Probability distribution of two voxel time series.

The results of activation analysis are obtained using the divergence estimation T_{ISE} (8) and independent t-test, respectively. Using the information from the analysis of dynamic phantom data, we also divide all the voxels into two groups, one lower range group (R_{IC} less than 3 percent) and another higher range group (R_{IC} greater than 3 percent). The activations of the lower and higher range group are shown in Figure 9.5 and 9.6, respectively.

The thresholds are set independently for the two methods. For a fair comparison, in both Figure 9.5 and 9.6 the top 200 voxels are shown with the highest nonparametric divergence and parametric t score. And as a constraint, activations from isolated single voxel are deleted in case of stochastic error. Comparing Figures 9.5 and 9.6, we find out that at the higher range the two

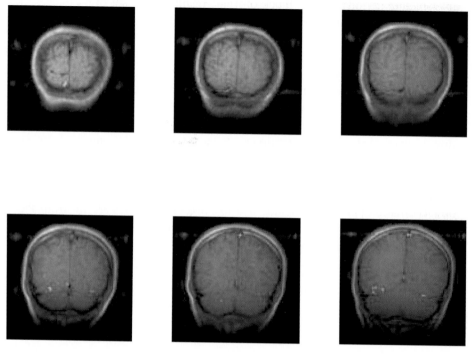

Figure 9.5. Activation map (intensity) at lower range of signal intensity change (below 3 percent). Red: distribution divergence measure based on mutual information T_{ISE} (8), Green: independent t statistics, and Yellow: activation overlapping by the two methods.

methods showed global similarity but local differences. However, at the lower range, the two methods did show very different activations, as can be inferred from the analysis of the dynamic phantom. Due to the deletion of activation from isolated single voxels, the parametric t-statistic (green) shows less activation than the divergence methods (red), while the latter give more concentrated activation map.

4.   Discussion

In this chapter a methodology for fMRI analysis based on divergence measures is evaluated. Results of single voxel based activation analysis of the dynamic phantom and human data are obtained using the two nonparametric divergence methods and compared with parametric t-test. The potential advantage of our divergence measure is that it can provide more information than those that use only second order statistics (such as cross correlation and t-test) and it does not require any assumption on the data distribution (e.g., Gaussian distribution). The analysis of the dynamic phantom data and the visual cortex

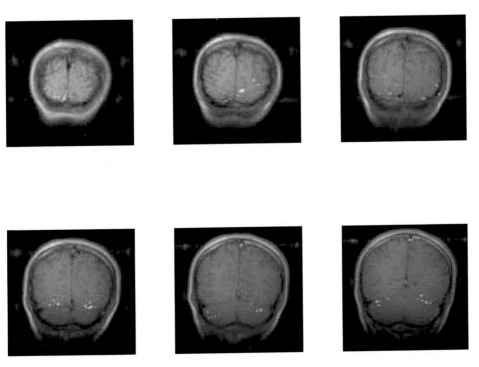

Figure 9.6. Activation (intensity) map at higher range of signal intensity changes (above 3 percent). Red: distribution divergence measure based on mutual information T_{ISE} (8), Green: independent t statistics, and Yellow: activation overlapping by the two methods.

study of human subject (Figures 9.5 and 9.6) showed the differences.

We note that a straight comparison between two detectors without ground truth is impossible, since any difference can have two potential causes: a missed detection by one method or false alarm of the other method. However, based on the study on the dynamic fMRI phantom data, we can conclude that the non-parametric divergences show better performance, both sensitivity and specificity when the signal intensity changes are small. This can be understood because the Gaussian assumption may not be justified when the R_{IC} is small, and the nonparametric divergence is totally independent of this assumption. Moreover, we did find in the visual cortex study that the differences between the activations maps of the two methods in Figures 9.4 and 9.6 primarily reside in the low intensity activity changes, and that overall one should trust more the activation map derived from the nonparametric divergences.

Although the nonparametric divergences show better performance than the parametric $t-$ statistics, the *specificity* of both methods is still not satisfying because the FPF is rather high in particular when the R_{IC} is low. One way to improve the estimation accuracy is to increase the sample size. Here we only have 64 data points for each voxel time series (32 "on" and 32 "off"), which is far from sufficient for a reliable estimation of the distribution divergence. This problem can be solved by increasing either the sampling rate (by using fast imaging schemes) or the time of experiment. Both methods will increase the data size and make it possible to describe the data distribution more accurately.

To increase further the realism of the simulation of "activation" pattern in our future work, we are considering to incorporate circuitry to simulate hemodynamics smoothing and delay, and add the cardiac and respiratory effects through periodic motion of the dynamic fMRI phantom in the scanner. Nevertheless, we consider the dynamic fMRI phantom approach superior to the digital manipulation of images because we preserve the intrinsic variability of the data collection.

5.   Summary

In this chapter we have developed a new information-theoretic approach to the calculation of fMRI activation maps. The many methods of calculation currently in use all depend, more or less, on *a priori* assumptions about either the analysis model (e.g, the linear model) or the data distributions (e.g., Gaussian distribution). The strength of the proposed divergence method is that it relies on solid theoretical principles and is easy to implement. And most importantly, while no assumption is made about the data distribution (non-parametric analysis), it can still uncover complex relationships (beyond second-order statistics). The dynamic fMRI phantom data clearly show the superiority of the nonparametric divergence approach compared to the parametric t-statistics, both in terms of sensitivity and selectivity at the low scale of intensity differences. Furthermore, the results of analyses of the human data are consistent with those from the dynamic phantom. We expect that the proposed nonparametric divergence methods will find important applications in cognitive neuroscience, given that researchers in this area study often focus on non-sensory-motor, anterior, and deep structures in the brain, where signal intensity changes are less marked and may will be undetected by traditional methods of analysis.

References

[1] Anderson, N.H. & Hall, P. (1994). Two-sample test statistics for measuring discrepancies between two multivariate probability density functions using kernel-based density estimates. *Journal of Multivariate Analysis.*, 50:41-54.

[2] Baillet, S. & Garnero, L. (1999). Combined MEG and EEG source imaging by minimization of mutual information. *IEEE Transactions on Biomedical Engineering*, 46(5):522-534.

[3] Bruning, J.L. & Kintz, B.L. (1987). *Computational handbook of statistics.*, Third edition, HarperCollins Publishers.

[4] Constable, R.T., Skudlarski,P., Mencl, E., Pugh, K.R., Fulbright, R.K., Lacadie, C., Shaywitz, S.E., & Shaywitz, B.A. (1998). Quantifying and comparing region-of-interest activation patterns in functional brain MR imaging: methodology considerations. *Magnetic Resonance Imaging*, 16(3):289-300.

[5] Diks, C., van Zwet, W.R., Takens, F., & DeGoede, J. (1996). Detecting differences between delay vector distributions. *Physical Review E*, 53(3):2169-2176.

[6] Frank, L.R., Buxton, R.B., & Wong, E.C. (1998). Probabilistic analysis of fMRI data. *Magnetic resonance in medicine*, 39:132-148.

[7] Friston, K.J., Holmes, A.P., Worsley, K.J., Poline, J-B., Frith, C.D., & Frackowiak, R.S.J. (1995). Statistical parametric maps in functional imaging: A general linear approach. *Human Brain Mapping*, 2:189-210.

[8] Glenberg, A.M. (1988). *Learning from data: An introduction to statistical reasoning.* Harcourt Brace Jovanovich, Publishers.

[9] Helstrom, C.W. (1968) *Statistical theory of signal detection.* 2nd edition, Pergamon Press Inc.

[10] Kapur, J.N., and Kesavan, H.K. (1992). *Entropy optimization principles with applications.* Academic Press, Inc.

[11] Kullback, S. (1968) *Information theory and statistics.* John Wiley & Sons, Inc.

[12] Lang, P.J., Bradley, M.M., Fitzsimmons, J.R., Cuthbert, B.N., Scott, J.D., Moulder, B., and Nangla, V. (1998) Emotional arousal and activation of the visual cortex: An fMRI analysis. *Psychophysiology,* 35:1-13.

[13] Lange, N. (1999). Statistical procedure for fMRI. Chapter 27 in *Functional MRI,* C.T.W.Moonen & P.A.Bandettini (Eds.), Springer-Verlag.

[14] Le, T.H. & Hu, X. (1997). Methods for accessing accuracy and reliability in functional MRI. *NMR in Biomedicine,* 10:160-164.

[15] Maes, F., Collignon, A., Vandermeulen, D., Marchal, G., & Suetens, P. (1997). Multimodality image registration by maximization of mutual information. *IEEE Transactions on Medical Imaging,* 16(2):187-198.

[16] Moonen, C.T.W. & Bandettini, P.A. (Eds.) (1999). *Functional MRI,* Springer-Verlag.

[17] Parzen, E. (1968). On estimation of a probability density function and mode. *Ann. Math. Stat.,* 33:1065-1076.

[18] Principe, J.C., Xu, D.X., and Fisher, J.W. (2000). Information-theoretic learning. Chapter in S. Haykin (eds.), *Unsupervised adaptive filtering, Volume I,* pp.265-319, John Wiley & Sons, Inc.

[19] Shannon, C.E. (1948). A mathematical theory of communication. *Bell System Tech. J.,* 27:379-423, 623-659, 1948.

[20] Sorenson, J.A. & Wang, X. (1996). ROC methods for evaluation of fMRI techniques. *Magn. Reson. Med.,* 36:737-744.

[21] Tsai, A., Fisher, J.W., Wible, C., Wells, W.M., Kim, J., & Willsky, A. (1999). Analysis of functional MRI data using mutual information. In *Second International Conference of Medical Image Computing and Computer-assisted Intervention,* 1679:473-480.

[22] Worsley, K.J., & Friston, K.J. (1995). Analysis of fMRI time-series revisited-again. *NeuroImage,* 2:173-181.

[23] Worsley, K.J., Marrett, S., Neelin, P., Vandal, A.C., Friston, K.J., and Evans, A.C. (1996). A unified statistical approach for determining significant signals in images of cerebral activations. *Human Brain Mapping,* 4:58-73.

[24] Zhao, Q., Principe, J.C., Bradley, M.M., & Lang, P.J. (2000). fMRI Analysis: Distribution divergence measure based on quadratic entropy. In *NeuroImage*, 11(5): 521, Part 2 of 2 parts.

Chapter 10

YEAST SAGE EXPRESSION LEVELS ARE RELATED TO CALCULATED MRNA FOLDING FREE ENERGIES

William Seffens
Department of Biological Sciences
Clark Atlanta University
Atlanta, GA 30314
wseffens@mediaone.net

Zarinah Hud
Department of Biological Sciences
Clark Atlanta University
Atlanta, GA 30314
zhud@cau.edu

David W. Digby
Department of Biological Sciences
Clark Atlanta University
Atlanta, GA 30314
ddigby@mindspring.com

Abstract   Free energies of folding for native mRNA sequences are more negative than calculated free energies of folding for randomized mRNA sequences with the same mononucleotide base composition and length. Randomization only of the coding region of most genes also yields folding free energies of less negative magnitude than those of the original mRNA sequences. For 79 mRNA sequences selected from a yeast SAGE library, the free energy minimization calculations of native mRNA sequences are also usually more negative than randomized mRNA sequences, as above. This difference can be expressed as a bias using standard deviation units. We also observed that if this yeast SAGE data is grouped according to expression levels, the mean folding free energy bias is different between the high, average, and low expression-level genes. A t-Test for paired two-samples of means shows a significant difference in folding free energies between high and

P.M. Pardalos and J. Principe (eds.), Biocomputing, 175-184.

low expression yeast genes. Thus the sequences of these yeast genes typically give rise to more stable secondary mRNA structures in high expression genes than in single-copy genes. The results of this study could serve as a foundation for comparison with other genomes, which in turn will allow investigating how the folding bias may be affected by specific characteristics of each organism, such as growth temperature, dinucleotide composition, or GC content of the genome.

Keywords: mRNA folding, SAGE, yeast genome

Introduction

Unlike the complementary strands of double helix DNA, RNA is found most frequently as a single stranded molecule. The capability to pair with a complementary strand also exists for RNA, but in most cases a reverse complementary molecule with which to pair is not available. Instead, a single molecule typically folds back upon itself, and pairing occurs between stretches of complementary bases along these reversed partial strands of RNA. The details of this process, and therefore the shape of the resulting structure and the extent of actual pairing, are dependent upon the sequence of the RNA molecule.

Although primarily an intermediate template along the way toward the synthesis of protein, messenger RNA can also fold in this way, but the functional consequences of such folding are less well understood. Several studies have demonstrated that mRNA stability may be an important factor in control of gene expression for certain genes (De Smit and van Duin, 1990; Wennborg *et al*, 1995). In some cases structural RNA features are thought to be involved in regulating the degradation of mRNA (Emory *et al*, 1992). RNA structural features may also influence control of gene expression. In the 5' region of some bacterial genes, for example, short stem-loop structures of folded RNA switch gene expression on and off (Love *et al*, 1988). It is highly likely, therefore, that the sequences of many genes have evolved in such a way as to allow particular folding patterns compatible with the functional properties of those genes. In the protein-encoding regions of a gene, this can still happen by selection from among the synonymous codons available for most amino acids, since this would have no impact upon the amino acid sequence. In a more speculative sense, these considerations could have influenced the evolutionary selection of the genetic code, or even the choice of nucleotide bases that occur in DNA and RNA (Digby and Seffens, 1999).

In silico folding of mRNAs

Free energy minimizing algorithms such as RNAstructure (Mathews et al) and Zuker's MFOLD program (Jacobson *et al*, 1998) are often utilized in predicting RNA secondary structures. The folding structures predicted typically consist of a family of structures that have the same or nearly the same free energy. A greater negative free energy indicates a more stable folding configuration. The thermodynamic treatment concerned in this work involves only the depth of the potential energy well represented by a folded mRNA molecule, rather than the actual secondary structures. Changes in the free energy of folding, expressed in the form of a standard deviation unit, has been called a segment score (Le and Maizel, 1989). The segment score is the difference between the free energy of a native sequence and the average free energy of a set of randomized sequences, divided by the standard deviation of the energies from the randomized set.

We have used RNAstructure v3.21 to investigate the folding characteristics of gene sequences from a variety of different organisms. Significant biases have been detected in the free energy of folding for native sequences, as compared to the average of several randomized versions of the same sequences (Seffens and Digby, 1999). This bias in genes is strongly dependent upon dinucleotide frequencies, which are known to be non-random, and changes with the source organisms examined (Workman and Krogh, 1999). If dinucleotide composition is preserved in the process of randomization, the observed bias in free energy is still present for rRNA, but is insignificant for mRNAs. Randomization subject only to preservation of mononucleotide composition reflects a first-order mutation model of DNA evolution (Conte *et al*, 1999). Double nucleotide substitutions that could preserve dinucleotide frequencies are less frequent by an order of magnitude (Averof *et al*, 2000). Therefore preserving mononucleotide frequencies in the randomized sets is more likely to model the most frequent mutation effects from which the native sequences are conserved.

Serial analysis of gene expression

Serial analysis of gene expression (SAGE), is an recent tool for studying the transcriptome, and provides a method of assessing the variation in the identity or amount of all genes expressed in different tissues at different times in development, or in normal versus abnormal cells (Velculescu, 1997). The data resulting from the SAGE technique is a list of tags, with their corresponding count values in the library, and thus is a digital representation of cellular gene expression. Tags are nucleotide sequences of a defined length typically 18 nucleotides for yeast libraries. One major advantage of using SAGE is that large numbers of expressed genes are analyzed at one time very quickly. It is possible to analyze up to one thousand transcripts in a few hours. With this technique, one can analyze the effects of drugs on tissue, identify disease related genes, and provide insight into disease and developmental pathways. In this work, yeast SAGE data was analyzed to find a possible correlation between SAGE expression levels and the mRNA folding bias.

Utilizing mRNA sequences from a Yeast SAGE library

Yeast SAGE data was downloaded via the World Wide Web at SAGEnet (www.sagenet.org) or NCBI-SAGEmap (www.ncbi.nlm.nih.gov) (Table 10.1). Seventy-nine mRNA sequences were selected from SAGEnet to sample a wide range of expression levels. The mRNA sequences selected were less than 3000 bases long due to limitations in the folding programs used. The number of times a specific tag is observed in the SAGE library reflects the expression level of the corresponding transcript (Velculescu, 1997).

Table 10.1.   Summary of Saccharomyces cerevisiae genome (from NCBI)

Chromosome Number	Total Bases	Protein Genes	RNA genes	Accession Number	Bases per gene
1	230203	107	4	NC001133	2074 to 2151
2	813140	427	16	NC001134	1836 to 1904
3	315339	172	14	NC001135	1695 to 1833
4	1531929	818	29	NC001136	1809 to 1872
5	576869	287	27	NC001137	1837 to 2010
6	270148	134	10	NC001138	1876 to 2016
7	1090936	571	42	NC001139	1780 to 1910
8	562639	284	13	NC001140	1894 to 1981
9	439885	220	11	NC001141	1904 to 1999
10	745440	387	29	NC001142	1792 to 1926
11	666445	335	20	NC001143	1877 to 1989
12	1078172	547	28	NC001144	1875 to 1971
13	924430	490	30	NC001145	1778 to 1886
14	784328	420	19	NC001146	1787 to 1867
15	1091283	572	30	NC001147	1813 to 1908
16	948061	499	23	NC001148	1816 to 1900
Totals:	12069247	6270	345		

Average 6270 protein-coding genes => 1925 bases per gene
Average for all 6615 total genes => 1825 bases per gene

All 79 mRNA sequences that were selected from Saccharomyces cerevisiae, plus their respective randomized sequences, were folded in silico, using the RNAstructure 3.2 program (Table 10.1). Current computer algorithms necessarily limit the length of mRNA sequences that can be folded, and most organisms include many genes that are too long to be folded this way. The average sequence length of a Saccharomyces cerevisiae gene is considerably less than 3000 bases (see Table 10.2), but many sequences in the SAGE database were longer than that and could not be analyzed. In addition, only the coding sequences were folded, not the complete mRNA since the SAGE database contained only the start and stop sites of the protein sequence. The folding of a sub-region like the CDS is correlated to the folding free energy of the whole mRNA, as shown below.

Table 10.2.   SAGE expression levels and folding bias statistic

Expression Group	SAGE levels	Mean bias	Standard deviation	Number of Genes	t-Test: two sample assuming equal variance
High	81-636	-1.642	1.669	20	<– t statistic = -2.720
Medium	13-80	-0.552	1.717	33	| t critical two
Low	1-12	-0.492	1.199	26	<– tail = 2.015
All	1-636	-0.808	1.609	79	

Folding of mRNA subregions

Although the folded yeast genes shown in Table 10.2 contained only the coding sequence, the results are generally comparable to folding the complete mRNA. Genes whose sequence is too long to be folded as a single unit can be divided into major segments and folded separately. An estimated value for the total free energy of each full length sequence can then be calculated from the values for the fragments. The following arguments apply to these calculations: Let the calculated folding free energy for the native mRNA molecules be expressed as ΔG_{native}. Similarly, let the various randomized mRNA folding free energies be expressed as ΔG_{random} for the complete sequence, and as $\Delta G_{CDSrandom}$, and $\Delta G_{UTRrandom}$ for the regional segments. Then also define a $\Delta\Delta G$ term:

$$\Delta\Delta G_{whole} = \Delta G_{native} - \Delta G_{random} \tag{1}$$

where $\Delta\Delta G_{whole}$ is the free energy contribution of the complete mRNA sequence to the free energy of folding the native mRNA. In a similar manner, one could calculate the contribution of a sequence from subsets or particular regions of the mRNA, so that

$$\Delta\Delta G_{CDS} = \Delta G_{native} - \Delta G_{CDSrandom} \tag{2}$$

and

$$\Delta\Delta G_{UTR} = \Delta G_{native} - \Delta G_{UTRrandom} \tag{3}$$

These two regional randomization expressions neglect the contribution of interactions between regions. A free energy term representing interactions between the regions should be subtracted from equation (2). Likewise a term of equal magnitude would be subtracted from equation (3) since the interactions are identical. Since the complete native mRNA is composed of subset 1 and subset 2 regions, then the sum of equation (2) and (3) would be:

$$\Delta\Delta G_{whole} = \Delta\Delta G_{CDS} + \Delta\Delta G_{UTR} + d \qquad (4)$$

where d is the intra-region free energy term that accounts for interactions between the two subsets. It is possible in this thermodynamic treatment to consider the contribution to d by interactions between more than two sub-regions.

A plot of the sum of $\Delta\Delta G_{CDS}$ and $\Delta\Delta G_{UTR}$ versus $\Delta\Delta G_{whole}$ is shown in Figure 10.1 using a set of mRNA sequences from Seffens and Digby (1999). A linear regression analysis of the data points from 51 mRNAs yields an equation of the form (y = mx + b):

$$(\Delta\Delta G_{CDS} + \Delta\Delta G_{UTR}) = (1.083) * \Delta\Delta G_{whole} - 0.708 \qquad (5)$$

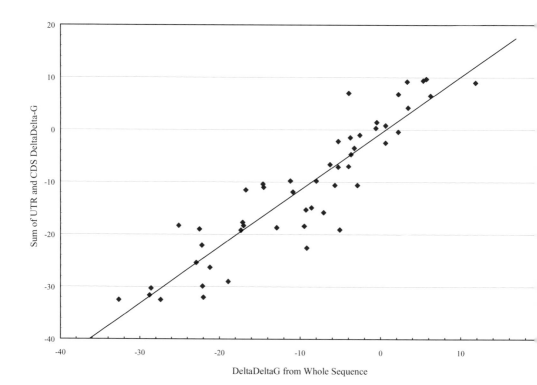

Figure 10.1.   Plot of $\Delta\Delta G_{CDS}$ plus $\Delta\Delta G_{UTR}$ versus $\Delta\Delta G_{whole}$

The computed regression coefficient is 0.914. This indicates that 84 percent of the variation in the sum of the regional free energy terms can be explained through the linear regression relationship in equation (5). The sign for the fitted intercept is negative, suggesting its value may represent the free energy contribution due to interactions between the UTRs and the CDS in the mRNA molecule, denoted by d. Contributions to total free energy of folding due to the 5'- and 3'-untranslated regions are smaller than the contribution due to the coding sequence.

Results and discussion

The free energy minimization calculations of native mRNA sequences from 79 mRNA sequences selected from yeast SAGE sequences are usually more negative than randomized mRNA sequences (Table 10.2). The average bias for the set is -0.808 in standard deviation units. A greater negative free energy indicates a more stable folding configuration, and folding free energies become more negative for increasing sequence length since more bonding interactions are possible on longer molecules. The results fitted to a linear line has a negative slope (Figure 10.2), but the regression coefficient is only 0.1 ($R \sup 2$). A t-Test for paired two-samples of means resulted in a significant difference with a t of P = .05 (44 degrees of freedom) equaling -2.720. This is significant since the t-statistic obtained from the data is larger than the 2.015 critical value for a two-tail test.

Although the biases calculated from mRNAs of different expression levels are rather small, they are significant at the 95 percent confidence level. These biases are calculated from an empirical molecular structure predictor, thereby suggesting that RNA secondary structures are important biological factors. The data set of yeast genes also shows a bias for base pairing within the CDS to be "in frame" (Seffens and Digby, 2000). These results support the use of folding free energies to classify genes (Seffens, 1999). It is unclear whether mRNA sequences have evolved to produce more stable RNA structures compared to mononucleotide randomized sequences, or as a result of biases in dinucleotide frequencies. A genetic algorithm program has found that mRNA folding is involved in selection of genetic codes subject to mutation (Digby and Seffens, 1999). This simulation studied the evolution of genetic codes in an ecosystem of protocells. It is possible that the bias in dinucleotide frequencies is due to the structure of the genetic code encouraging formation of RNA structures. An argument based on graph theory has been presented suggesting that "in frame" base-pairing is related to the folding free energy biases (Seffens and Digby, 2000).

The results of this study suggest a larger project of wider interest. If the biases for all of the mRNAs of an organism were to be computed, a mean value could be

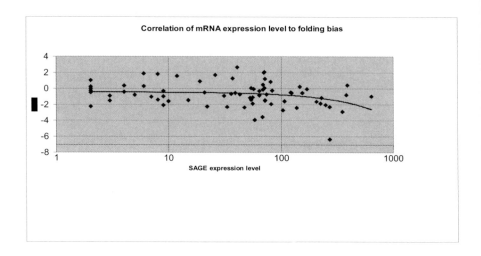

Figure 10.2.   Scatter plot of mRNA folding bias correlated to SAGE expression levels. Linear line fitted to data has negative slope. Note log scale for mRNA expression levels.

used to characterize the whole genome. The results of such a study could serve as a foundation for comparison with other genomes, which in turn would allow investigating how the folding bias may be affected by specific characteristics of each organism, such as growth temperature, dinucleotide composition, or GC content of the genome. There are 6270 protein coding genes identified by the NCBI genome listing for Saccharomyces cerevisiae, plus 345 structural RNA genes, all contained within 12,069,247 nucleotide bases (Table 10.1). In previous work, an average 2000 base gene could be folded in 10 minutes, or 6 per hour. Therefore to fold that gene 11 times (native + 10 shuffled) would take 110 minutes, or 1.83 hours. Given 6270 genes, this is 11,495 hours, or 479 days, a large, but not impossible task.

Acknowledgments

This work was supported (or partially supported) by NIH grant GM08247, by a Research Centers in Minority Institutions award, G12RR03062, from the Division of Research Resources, National Institutes of Health, and NSF CREST Center for Theoretical Studies of Physical Systems (CTSPS) Cooperative Agreement #HRD-9632844.

References

[1] Averof, M, Rokas, A, Wolfe, K., and Sharp, P. (2000) "Evidence for a high frequency of simultaneous double-nucleotide substitutions", Science 287:1283-1286.

[2] Conte, L., Chothia, C, and Janin, J. (1999) J. Mol. Biol. 285:2177.

[3] De Smit, M.H., and van Duin, J. (1990) Prog. Nucleic Acid Res. Mol. Biol. 38:1-35.

[4] Digby, D. and Seffens, W. (1999), Evolutionary Algorithm Analysis of the Biological Genetic Codes, Proceedings of 1999 Genetic and Evolutionary Computation Conference, Orlando, FL, p 1440.

[5] Emory, S.A., Bouvet, P., and Belasco, J.G. (1992). A 5'-terminal stem-loop structure can stabilize mRNA in Escherichia coli. Genes Dev. 6:135-148.

[6] Jacobson, A.B., Arora, R., Zuker, M., Priano, C., Lin, C.H., and Mills, D.R., (1998). J. Mol. Biol. 274:589-600.

[7] Le, S.Y. and Maizel, J.V. Jr. (1989). A method for assessing the statistical significance of RNA folding. J. Theor.Biol. 138:495-510.

[8] Love Jr., H.D., Allen-Nash, A., Zhao, Q., and Bannon, G.A. (1988). mRNA stability plays a major role in regulating the temperature-specific expression of a Tetrahymena thermophila surface protein. Mol. Cell. Biol. 8:427-432.

[9] Mathews, D.H., Andre, T.C., Kim, J., Turner, D.H., and Zuker, M. (1998) In Leontis, N.B., and SantaLucia, J., Jr (eds), Molecular Modeling of Nucleic Acids. American Chemical Society Symposium Series 682, Washington, DC, pp. 246-257.

[10] D.H. Mathews, J. Sabina, M. Zuker, and D. H. Turner. Expanded Sequence Dependence of Thermodynamic Parameters Improves Prediction of RNA Secondary Structure. Journal of Molecular Biology, In Press.

[11] W. Seffens 1999. mRNA classification based on calculated folding free energies, WWW Journal of Biology (http://epress.com/w3jbio/), vol.4-3.

[12] Seffens, W., and Digby, D. (1999). MRNAs have greater negative folding free energies than shuffled or codon choice randomized sequences. Nuc. Acid Res. 27:1578-1584.

[13] Seffens, W. and Digby, D., (2000) Gene Sequences are Locally Optimized for Global mRNA Folding, Optimization in Computational Chemistry and Molecular Biology, (C. Floudas and P. Pardalos, eds.) Kluwer Academic Press. pp 131-140.

[14] Velculescu, V.E., (1997) Cell, 88,243.

[15] Wennborg, A., Sohlberg, B., Angerer, D., Klein, G., and Von Gabain, A. (1995). A human RNase E-like activity that cleaves RNA sequences involved in mRNA stability control. Proc. Natl. Acad. Sci. 92:7322-7326.

[16] Workman, C., and Krogh, Anders (1999). No evidence that mRNAs have lower folding free energies than random sequences with the same dinucleotide distribution. Nuc. Acid Res. 27:4816-4822.

Chapter 11

SOURCES AND SINKS IN MEDICAL IMAGE ANALYSIS

Kaleem Siddiqi

School of Computer Science &

Centre for Intelligent Machines

McGill University

3480 University Street

Montréal QC H3A 2A7, Canada

siddiqi@cim.mcgill.ca

Abstract   Vector fields which arise in image analysis are often assumed to be conservative or divergence free. Such vector fields arise, for example, when considering the velocity of an incompressible fluid. Conversely, locations where energy is not conserved are sources or sinks. In this paper we show how sources and sinks connect two disparate problems of interest in medical image analysis: the computation of skeletons and the segmentation of tubular structures such as blood vessels. Using the divergence theorem, we review algorithms for solving both problems and present numerical results.

Keywords: Divergence, vector fields, skeletons, vessel segmentation

1.   Introduction

The 2D skeleton of a closed set $A \subset \mathcal{R}^2$ is the locus of centers of maximal open discs contained within the complement of the set ([2]). An open disc is maximal if there exists no other open disc contained in the complement of A that properly contains the disc. The 3D skeleton of a closed set $A \subset \mathcal{R}^3$ is defined in an analogous fashion as the locus of centers of maximal open spheres contained in the complement of the set. Both types of representations are of significant interest for a number of applications in biomedicine, including object representation ([10, 15]), registration ([7]) and segmentation ([13]). Such descriptions provide a compact representation while preserving the object's

P.M. Pardalos and J. Principe (eds.), Biocomputing, 185-198.

genus and retain sufficient local information to reconstruct it. This facilitates a number of important tasks including the quantification of the local width of a complex structure, e.g., that of the grey matter in the human brain, and the analysis of its topology. Despite their popularity, the stable numerical computation of skeletons remains a challenge, particularly in 3D.

A seemingly unrelated problem, which is of importance in image-guided neuro-surgery, is the segmentation of tubular structures such as blood vessels in magnetic resonance angiography (MRA) images. Proposed approaches include those which characterize the physical properties of blood flow ([17]) as well as ones which exploit properties of the Hessian to obtain geometric estimates ([6]). A multi-scale method for the detection of curvilinear structures in 2D and 3D data has been introduced by [5] and a method for obtaining 3D vascular trees which calculates vessel centerlines as intensity ridges has been introduced by [4]. More recently, a co-dimension 2 geometric flow (a curve evolving in 3D) has been introduced by [8], which is based on the results of [1].

In this paper we show that there is in fact a close link between these two problems, which arises by considering the energy associated with an appropriate vector field for each. The sources and sinks of the vector field turn out to be key, and these can be robustly estimated using an application of the divergence theorem. We review the theory and present algorithms for solving both problems, along with numerical results.

2.   Divergence-based skeletons

The first problem we consider is the computation of skeletons in 2D and 3D. Here we shall exploit the fact that the locus of skeletal points coincides with the singularities of a Euclidean distance function from the boundary of an object.

Let $\mathcal{C}$ be the vector of coordinates of a closed curve $(x(p, t), y(p, t))$ where p parametrizes the curve and t is an independent parameter which denotes its evolution in time to give a family of curves. Consider the grassfire flow ([2])

$$\frac{\partial \mathcal{C}}{\partial t} = \mathcal{N} \tag{1}$$

such that each point on its boundary is moving with unit speed in the direction of the inward normal $\mathcal{N}$. In recent work, we have shown that this formulation leads to a Hamilton-Jacobi equation on the Euclidean distance function ([14]). Specifically, let D be the Euclidean distance function to the initial curve $\mathcal{C}_0$. The magnitude of its gradient, $\|\nabla D\|$, is identical to 1 in its smooth regime. With $\mathbf{q} = (x, y)$, $\mathbf{p} = (D_x, D_y)$, the Hamiltonian system is given by

$$\dot{\mathbf{p}} = -\frac{\partial H}{\partial \mathbf{q}} = (0, 0), \qquad \dot{\mathbf{q}} = \frac{\partial H}{\partial \mathbf{p}} = -(D_x, D_y), \tag{2}$$

with an associated Hamiltonian function $H = 1 - (D_x^2 + D_y^2)^{\frac{1}{2}}$. It is straightforward to show that all Hamiltonian systems are conservative and hence divergence free. Conversely, when trajectories of the system intersect, a conservation of energy principle is violated. This suggests a natural approach for computing the skeleton: compute the divergence of the gradient vector field $\dot{\mathbf{q}}$ and detect locations where it is not zero. The divergence is defined as the net outward flux per unit area, as the area about the point shrinks to zero:

$$\mathrm{div}(\dot{\mathbf{q}}) \equiv \lim_{\Delta a \to 0} \frac{\int_L < \dot{\mathbf{q}}, \mathcal{N} > \mathrm{dl}}{\Delta a} \tag{3}$$

Here Δa is the area, L is its bounding contour and $\mathcal{N}$ is the outward normal at each point on the contour. Via the divergence theorem

$$\int_a \mathrm{div}(\dot{\mathbf{q}})\mathrm{da} \equiv \int_L < \dot{\mathbf{q}}, \mathcal{N} > \mathrm{dl}. \tag{4}$$

In other words, the integral of the divergence of the vector field within a finite area gives the net outward flux through the contour which bounds it. Locations where the flux is negative, and hence energy is lost, correspond to sinks or skeletal points of the interior. Similarly, locations where the flux is positive correspond to sources or skeletal points of the exterior. This is an integral formulation and is therefore numerically more robust than the standard computation of divergence as the sum of partial derivatives in the vector field's component directions. The extension of the analysis to 3D is straightforward: simply replace the initial closed curve $\mathcal{C}$ with a closed surface $\mathcal{S}$ in Eq. (1), add a third coordinate z to the phase space in Eq. (2), and replace the area element with a volume element and the contour integral with a surface integral in Eq. (3).

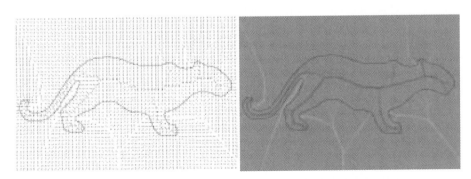

Figure 11.1. Left: The gradient vector field of a signed Euclidean distance function D to the boundary of a panther shape. Right: The associated total outward flux. Whereas the smooth regime of the vector field gives zero flux (medium gray), its sinks correspond to the skeleton of the interior (dark gray) and its sources to the skeleton of the exterior (light gray).

Figure 11.1 illustrates the divergence-based computation on the silhouette of a panther shape. The gradient vector field of the Euclidean distance function is shown on the left, with the outward flux on the right. The smooth regime of the vector field gives zero flux (medium gray), its sinks coincide with the medial axis of the interior (dark gray), and its sources with the medial axis of the exterior (light gray). Although a threshold on the flux map yields a close approximation to the skeleton, further criteria must be introduced in order to guarantee that the result is homotopic to the original shape and is thin (has no interior). The main idea is to characterize digital points which can be removed without altering the object's topology. The object is then thinned by deleting removable points sequentially, in order of their flux values, while fixing or anchoring endpoints. The process converges when all remaining digital endpoints are either endpoints or are not removable.

The characterization of a removable point in 2D is straightforward. Consider the 3x3 neighborhood of a digital point P and construct a neighborhood graph by placing edges between all pairs of neighbors (not including P) that are 4-adjacent or 8-adjacent to one another. If there are any cycles of length 3 remove the diagonal edges. P is removable if and only if the resulting graph is a tree. It is also easy to see that an endpoint in 2D is one which has a single neighbor in a 3x3 neighborhood, or has two neighbors both of which are 4-adjacent to one another. In 3D a criterion for removable points is more subtle and is based on the results of [9]. Let C^* be the number of 26-connected components 26-adjacent to P in $O \cap N_{26}^*$, where O is the object and N_{26}^* is the 26-neighborhood with P removed. Let $\bar{C}$ be the number of 6-connected components 6-adjacent to P in $\bar{O} \cap N_{18}$, where N_{18} is its 18-neighborhood. P is removable if $C^* = \bar{C} = 1$. Endpoint criteria in 3D are also more difficult to develop. One may restrict the definition to a 6-connected neighborhood, in which case an endpoint may be viewed as the end of a 6-connected curve or a corner or a point on the rim of a 6-connected surface ([3]) or one may examine nine digital planes which intersect at the point ([12]).

The thinning process can be made very efficient by using a heap. A full description of the procedure is given in ([3]); its worst-case complexity can be shown to be $\mathcal{O}(n) + \mathcal{O}(k\log(k))$, where n is the total number of points in the 2D or 3D array and k is the number of points in the interior of the object. Numerical results are presented in Figure 11.2 for a volumetric data set consisting of brain ventricles segmented from a magnetic resonance (MR) image. The figure illustrates the effect of thresholding the flux map (second column) where the skeletons corresponding to the views in the first column are accurate, but have holes. With the same threshold the divergence-based thinning algorithm yields a thin structure which preserves the object's topology, Figure 11.2 (third column). The ventricles reconstructed from the medial surfaces in the fourth column are shown in the fifth column.

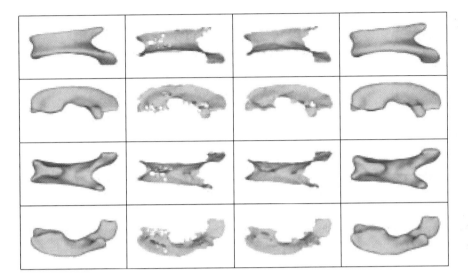

Figure 11.2. FIRST COLUMN: Four views of the ventricles of a brain, segmented from volu-metric MR data using an active surface. SECOND COLUMN: The corresponding medial surfaces obtained by thresholding the divergence map. THIRD COLUMN: The divergence-based medial surfaces obtained using the same threshold, but with the incorporation of homotopy preserving thinning. FOURTH COLUMN: The ventricles reconstructed from the divergence-based medial surfaces in the previous column.

3. Flux maximizing flows

We now consider the second problem, that of segmenting elongated structures such as blood vessels in MRA images. Brightness in MRA is proportional to the magnitude of blood flow velocity. This leads to the constraint that in their vicinity, the gradient vector field of an MRA image should be locally orthogonal to blood vessel boundaries. Thus, a natural principle to use towards the recovery of these boundaries is to maximize the flux of the gradient vector field through an evolving curve (in 2D) or surface (in 3D). We now derive this flux maximizing flow.

As before, let $\mathcal{C} = \mathcal{C}(x(p,t), y(p,t))$ be a smooth family of closed curves evolving in the plane. Here t parametrizes the family and p the given curve. Without loss of generality we shall assume that $0 \le p \le 1$, i.e., that $\mathcal{C}(0,t) = \mathcal{C}(1,t)$. We shall also assume that the first derivatives exist and that $\mathcal{C}'(0,t) = \mathcal{C}'(1,t)$. The unit tangent $\mathcal{T}$ and the unit inward normal $\mathcal{N}$ to $\mathcal{C}$ are given by

$$\mathcal{T} = \frac{\begin{pmatrix} x_p \\ y_p \end{pmatrix}}{\|\mathcal{C}_p\|} = \begin{pmatrix} x_s \\ y_s \end{pmatrix} ; \mathcal{N} = \frac{\begin{pmatrix} -y_p \\ x_p \end{pmatrix}}{\|\mathcal{C}_p\|} = \begin{pmatrix} -y_s \\ x_s \end{pmatrix}$$

where s is the arc-length parametrization of the curve. Now consider a vector field $\mathcal{V} = (V_1(x, y), V_2(x, y))$ defined for each point (x, y) in $\mathcal{R}^2$. The total inward flux of the vector field through the curve is given by the contour integral

$$Flux(t) = \int_0^1 \langle \mathcal{V}, \mathcal{N} \rangle \, \|\mathcal{C}_p\| \, dp = \int_0^{L(t)} \langle \mathcal{V}, \mathcal{N} \rangle \, ds$$

where $L(t)$ is the Euclidean length of the curve. The circulation of the vector field along the curve is defined in an analogous fashion as

$$Circ(t) = \int_0^1 \langle \mathcal{V}, \mathcal{T} \rangle \, \|\mathcal{C}_p\| \, dp = \int_0^{L(t)} \langle \mathcal{V}, \mathcal{T} \rangle \, ds.$$

It can be shown that the gradient flow which maximizes the rate of increase of the total inward flux is obtained by moving each point of the curve in the direction of the inward normal by an amount proportional to the divergence of the vector field:

Theorem 1 *The direction in which the inward flux of the vector field $\mathcal{V}$ through the curve $\mathcal{C}$ is increasing most rapidly is given by $\frac{\partial \mathcal{C}}{\partial t} = div(\mathcal{V})\mathcal{N}$.*

Proof: Define the perpendicular to a vector $\mathcal{W} = (a, b)$ as $\mathcal{W}^\perp = (-b, a)$. The following properties hold:

$$\left\langle \mathcal{U}, \mathcal{W}^\perp \right\rangle = -\left\langle \mathcal{U}^\perp, \mathcal{W} \right\rangle$$
$$\left\langle \mathcal{U}^\perp, \mathcal{W}^\perp \right\rangle = \langle \mathcal{U}, \mathcal{W} \rangle. \tag{5}$$

We now compute the first variation of the flux functional with respect to t

$$Flux'(t) = \underbrace{\int_0^{L(t)} \langle \mathcal{V}_t, \mathcal{N} \rangle \, ds}_{I_1} + \underbrace{\int_0^{L(t)} \langle \mathcal{V}, \mathcal{N}_t \rangle \, ds}_{I_2}.$$

With

$$\mathcal{V}_t = \left(\frac{\partial V_1}{\partial t}, \frac{\partial V_2}{\partial t} \right) = (\langle \nabla V_1, \mathcal{C}_t \rangle, \langle \nabla V_2, \mathcal{C}_t \rangle)$$

$$I_1 = \int_0^{L(t)} \langle \mathcal{C}_t, x_s \nabla V_2 - y_s \nabla V_1 \rangle \, ds.$$

For I_2 we first switch to parametrization by p:

$$I_2 = \int_0^{L(t)} \langle \mathcal{V}, \mathcal{N}_t \rangle \, ds = \int_0^1 \left\langle \mathcal{V}, \left(\begin{array}{c} -y_{pt} \\ x_{pt} \end{array} \right) \right\rangle \, dp.$$

Now, using integration by parts

$$I_2 = \underbrace{\left\langle \mathcal{V}, \begin{pmatrix} -y_t \\ x_t \end{pmatrix} \right\rangle \Big]_0^1}_{\text{equals } 0} - \int_0^1 \left\langle \begin{pmatrix} -y_t \\ x_t \end{pmatrix}, \mathcal{V}_p \right\rangle dp$$

Using the properties of scalar products in Eq. (5) and the fact that

$$\mathcal{V}_p = \left(\frac{\partial V_1}{\partial p}, \frac{\partial V_2}{\partial p} \right) = \left(\langle \nabla V_1, \mathcal{C}_p \rangle, \langle \nabla V_2, \mathcal{C}_p \rangle \right)$$

we can rewrite I_2 as follows

$$I_2 = \int_0^1 \left\langle \begin{pmatrix} x_t \\ y_t \end{pmatrix}, \mathcal{V}_p^\perp \right\rangle dp$$

$$= \int_0^1 \left\langle \mathcal{C}_t, \begin{pmatrix} -\langle \nabla V_2, \mathcal{C}_p \rangle \\ \langle \nabla V_1, \mathcal{C}_p \rangle \end{pmatrix} \right\rangle dp.$$

Switching to arc-length parametrization

$$I_2 = \int_0^{L(t)} \left\langle \mathcal{C}_t, \begin{pmatrix} -\langle \nabla V_2, \mathcal{T} \rangle \\ \langle \nabla V_1, \mathcal{T} \rangle \end{pmatrix} \right\rangle ds.$$

Combining I_1 and I_2, the first variation of the flux is

$$\int_0^{L(t)} \left\langle \mathcal{C}_t, x_s \nabla V_2 - y_s \nabla V_1 + \begin{pmatrix} -\langle \nabla V_2, \mathcal{T} \rangle \\ \langle \nabla V_1, \mathcal{T} \rangle \end{pmatrix} \right\rangle ds.$$

Thus, for the flux to increase as fast as possible, the two vectors should be made parallel:

$$\mathcal{C}_t = x_s \nabla V_2 - y_s \nabla V_1 + \begin{pmatrix} -\langle \nabla V_2, \mathcal{T} \rangle \\ \langle \nabla V_1, \mathcal{T} \rangle \end{pmatrix}.$$

Decomposing the above three vectors in the Frenet frame $\{\mathcal{T}, \mathcal{N}\}$, dropping the tangential terms (which affect only the parametrization of the curve) and making use of the properties of scalar products

$$\mathcal{C}_t = \{ x_s \langle \nabla V_2, \mathcal{N} \rangle - y_s \langle \nabla V_1, \mathcal{N} \rangle$$
$$+ \left\langle \begin{pmatrix} -\langle \nabla V_2^\perp, \mathcal{N} \rangle \\ \langle \nabla V_1^\perp, \mathcal{N} \rangle \end{pmatrix}, \mathcal{N} \right\rangle \} \mathcal{N}$$

Expanding all terms in the above equation

$$\mathcal{C}_t = (x_s(-V_{2x}.y_s + V_{2y}.x_s) - y_s(-V_{1x}.y_s + V_{1y}.x_s) +$$
$$\left\langle \left(\begin{array}{c} -V_{2y}.y_s - V_{2x}.x_s \\ V_{1y}.y_s + V_{1x}.x_s \end{array} \right), \mathcal{N} \right\rangle)\mathcal{N}$$
$$= (-V_{2x}.x_s.y_s + V_{2y}.x_s^2 + V_{1x}.y_s^2 - V_{1y}.x_s.y_s$$
$$+V_{2y}.y_s^2 + V_{2x}.x_s.y_s + V_{1y}.x_s.y_s + V_{1x}.x_s^2)\mathcal{N}$$
$$= (V_{1x}(x_s^2 + y_s^2) + V_{2y}(x_s^2 + y_s^2))\mathcal{N}$$
$$= (V_{1x} + V_{2y})\mathcal{N} = div(V).\mathcal{N}. \tag{6}$$

As a corollary to Theorem 1, we have

Corollary 1 *The direction in which the circulation of of the vector field V along the curve $\mathcal{C}$ is increasing most rapidly is given by $\frac{\partial \mathcal{C}}{\partial t} = div(V^\perp)\mathcal{N}$.*

Proof: Using the properties of scalar products in Eq. (5)

$$Circ(t) = \int_0^{L(t)} \langle V, \mathcal{T} \rangle \, ds$$
$$= \int_0^{L(t)} \left\langle V^\perp, \mathcal{T}^\perp \right\rangle \, ds$$
$$= \int_0^{L(t)} \left\langle V^\perp, \mathcal{N} \right\rangle \, ds$$

Hence the circulation of the vector field V along the curve is just the inward flux of the vector field $V^\perp$ through it and the result follows from Theorem 1.

It should not come up as a surprise that the flux maximizing flow can be extended to 3D. However, to show this we first need to set up some notation. Let $\mathcal{S} : [0, 1] \times [0, 1] \to \mathcal{R}^3$ denote a compact embedded surface with (local) coordinates (u, v). Let $\mathcal{N}$ be the inward unit normal. We set

$$\mathcal{S}_u := \frac{\partial \mathcal{S}}{\partial u}, \quad \mathcal{S}_v := \frac{\partial \mathcal{S}}{\partial v}.$$

Then the infinitesimal area on $\mathcal{S}$ is given by

$$d\mathcal{S} = (\|\mathcal{S}_u\|^2\|\mathcal{S}_v\|^2 - \langle \mathcal{S}_u, \mathcal{S}_v \rangle^2)^{1/2} du dv.$$

Let $V = (V_1(x, y, z), V_2(x, y, z), V_3(x, y, z))$ be a vector field defined for each point (x, y, z) in $\mathcal{R}^3$. The total inward flux of the vector field through the surface is defined in a manner analogous to the 2D case and is given by the surface integral

$$Flux(t) = \int_0^1 \int_0^1 \langle V, \mathcal{N} \rangle \, d\mathcal{S}. \tag{7}$$

However, an analogous definition does not exist for the circulation because the notion of a tangent in 2D is now replaced by that of a tangential plane [1]. It turns out that the divergence of the vector field is once again the key to maximizing the rate of increase of inward flux:

Theorem 2 *The direction in which the inward flux of the vector field V through the surface S is increasing most rapidly is given by $\frac{\partial S}{\partial t} = div(V)\mathcal{N}$.*

Sketch of Proof: As before, the idea is to calculate the first variation of the flux functional with respect to t:

$$Flux'(t) = \underbrace{\int_0^1 \int_0^1 \langle V_t, \mathcal{N} \rangle \, dS}_{I_1} + \underbrace{\int_0^1 \int_0^1 \langle V, \mathcal{N}_t \rangle \, dS}_{I_2}.$$

With

$$\mathcal{N} = \frac{S_u \bigwedge S_v}{\|S_u \bigwedge S_v\|} = \begin{pmatrix} N_1 \\ N_2 \\ N_3 \end{pmatrix}$$

$$I_1 = \int_0^1 \int_0^1 \langle S_t, (N_1 \nabla V_1 + N_2 \nabla V_2 + N_3 \nabla V_3) \rangle \, dS,$$

which has the desired form of an inner product of S_t with another vector. The basic idea behind calculating I_2 is to once again use integration by parts, but now on a double integral, along with the chain rule and properties of cross products. The expression can then be simplified and combined with I_1 to yield the gradient flow which maximizes the rate of increase of flux. The components of this flow in the tangential plane to the surface at each point can be dropped, since these affect only the parametrization. The flow then works out to be

$$S_t = div(V)\mathcal{N}. \tag{8}$$

Whereas thus far the development has remained abstract, the flux maximizing flow can be tailored to the problem of blood vessel segmentation being considered here. To see this, simply let the gradient $\nabla \mathbf{I}$ of the original intensity image $\mathbf{I}$ be the vector field V whose inward flux through the evolving curve (or surface) is maximized. As with the computation of divergence-based skeletons, the numerical computation can be made robust by using the divergence theorem in Eq. (4). In our implementations we use this outward flux formulation along the boundaries of circles (in 2D) or spheres (in 3D) of varying radii, corresponding to a range of blood vessel widths. The chosen flux value at a particular location is the maximum (magnitude) flux over the range of radii. In contrast to other multi-scale approaches where combining information across scales is non-trivial ([6]) normalization across scales is straightforward in our

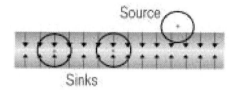

Figure 11.3.   An illustration of the gradient vector field in the vicinity of a blood vessel. As-suming a uniform background intensity, at its centerline, at the scale of the vessel's width, the total outward flux, which is proportional to the divergence, is negative (a sink). Outside the vessel, at a smaller scale, the total outward flux is positive (a source).

case. One simply has to divide by the number of entries in the discrete sum that approximates Eq. (4). Locations where the total outward flux (which is proportional to the divergence) is negative correspond to sinks and locations where the total outward flux is positive correspond to sources, as illustrated in Figure 11.3. Both sources and sinks play a crucial role: when seeds are placed within blood vessels, the sinks force them to evolve towards the blood vessel boundaries while the sources outside prevent the flow from leaking.

In order to implement the flow, we use the level set representation for curves flowing according to functions of curvature, due to Osher and Sethian ([11]). Let $\mathcal{C}(p,t) : S^1 \times [0, \tau) \to \mathbf{R}^2$ be a family of curves satisfying the curve evolution equation

$$\mathcal{C}_t = F\mathcal{N},$$

where F is an arbitrary (local) scalar speed function. Then it can be shown that if $\mathcal{C}(p, t)$ is represented by the zero level set of a smooth and Lipschitz continuous function $\Psi : \mathbf{R}^2 \times [0, \tau) \to \mathbf{R}$, the evolving surface satisfies

$$\Psi_t = F \, \|\nabla \Psi\| \, .$$

This last equation is solved using a combination of straightforward discretiza-tion and numerical techniques derived from hyperbolic conservation laws ([11]). For hyperbolic terms, care must be taken to implement derivatives with upwinding in the proper direction. The evolving curve $\mathcal{C}$ is then obtained as the zero level set of Ψ. The formulation is analogous for the case of surfaces evolving in 3D.

Figure 11.4 illustrates the flow of Eq. (6) on a portion of a 2D retinal an-giogram. Notice how a single seed evolves along the direction of shading (orthogonal to the image intensity gradient) to reconstruct thin or low contrast blood vessel boundaries. Figure 11.5 illustrates the flow of Eq. (8) on a 3D MRA image of blood vessels in the head. A maximal intensity projection of the data is shown on the top left, followed by the evolution of a few 3D spheres. These spheres were placed uniformly in regions of high flux, which is similar to the idea of using "bubbles" ([16]). Notice how the spheres elongate in the

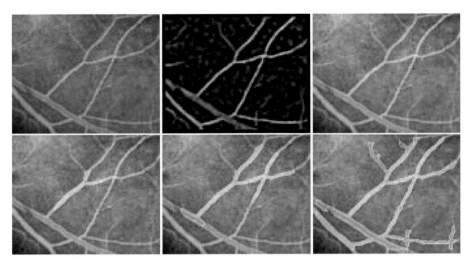

Figure 11.4. An illustration of the flux maximizing flow for a portion of a 2D retinal angiography image. The original is shown on the top left. The multi-scale flux (divergence) of the gradient vector field is shown on its right. Bright regions correspond to negative flux. The other images depict the evolution of a single seed, which follows the direction of shading to reconstruct the blood vessel boundaries.

direction of blood vessels, which is the expected behavior since it maximizes the rate of increase of inward flux through them. The main blood vessels, which have the higher flux, are the first to be captured.

4. Conclusions

The link between the computation of skeletons and the segmentation of blood vessels in medical imaging is provided by locating the sources and sinks of an appropriate vector field. On the one hand, sinks of the signed Euclidean distance function from the boundary of an object coincide with skeletal points of its interior and sources with skeletal points of its exterior. On the otherhand, the fastest way to maximize the inward flux of the gradient vector field of an MRA image through an evolving curve or surface is to move each point in the direction of the inward normal, by an amount proportional to the divergence. Sinks within blood vessel boundaries push the curve or surface towards them and sources outside prevent the flow from leaking. An important aspect of the numerical simulations presented here is the use of the divergence theorem to estimate the outward flux as an integral. This is numerically more robust than the standard computation as the sum of partial derivatives in the vector field's component directions. In fact, the latter definition breaks down at sources and sinks where the vector field is singular, and these are precisely the locations of interest for these two problems.

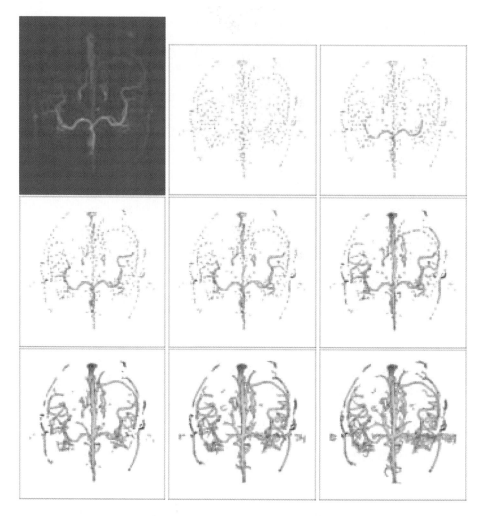

Figure 11.5. An illustration of the flux maximizing flow for a 3D MRA image. A maximum-intensity projection of the data is shown on the top left and the other images depict the evolution of a few isolated spheres. The main vessels, which have higher flux, are the first to be reconstructed.

Acknowledgments

I am grateful to Sylvain Bouix, Allen Tannenbaum, Alexander Vasilevskiy and Steve Zucker for collaborations on the results presented here.

References

[1] Ambrosio, L. and Soner, H. M. (1996). Level set approach to mean curvature flow in arbitrary codimension. *Journal of Differential Geometry*, 43:693–737.

[2] Blum, H. (1973). Biological shape and visual science. *Journal of Theoretical Biology*, 38:205–287.

[3] Bouix, S. and Siddiqi, K. (2000). Divergence-based medial surfaces. In *ECCV'2000*, pages 603–618, Dublin, Ireland.

[4] Bullitt, E., Aylward, S., Liu, A., Stone, J., Mukherjee, S. K., Coffey, C., Gerig, G., and Pizer, S. M. (1999). 3d graph description of the intracerebral vasculature from segmented mra and tests of accuracy by comparison with x-ray angiograms. In *IPMI'99*, pages 308–321.

[5] Koller, T. M., Gerig, G., Székely, G., and Dettwiler, D. (1995). Multiscale detection of curvilinear structures in 2-d and 3-d image data. In *ICCV'95*, pages 864–869.

[6] Krissian, K., Malandain, G., Ayache, N., Vaillant, R., and Trousset, Y. (1998). Model-based multiscale detection of 3d vessels. In *CVPR'98*, pages 722–727.

[7] Liu, A., Bullitt, E., and Pizer, S. M. (1998). 3d/2d registration via skeletal near projective invariance in tubular objects. In *MICCAI'98*, pages 952–963.

[8] Lorigo, L. M., Faugeras, O., Grimson, E. L., Keriven, R., Kikinis, R., Nabavi, A., and Westin, C.-F. (2000). Codimension-two geodesic active contours for the segmentation of tubular structures. In *CVPR'2000*, volume 1, pages 444–451.

[9] Malandain, G., Bertrand, G., and Ayache, N. (1993). Topological segmentation of discrete surfaces. *Int. Journal of Computer Vision*, 10(2):183–197.

[10] Näf, M., Kübler, O., Kikinis, R., Shenton, M. E., and Székely, G. (1996). Characterization and recognition of 3d organ shape in medical image analysis using skeletonization. In *IEEE Workshop on Mathematical Methods in Biomedical Image Analysis*.

[11] Osher, S. J. and Sethian, J. A. (1988). Fronts propagating with curvature dependent speed: Algorithms based on hamilton-jacobi formulations. *Journal of Computational Physics*, 79:12–49.

[12] Pudney, C. (1998). Distance-ordered homotopic thinning: A skeletonization algorithm for 3d digital images. *Computer Vision and Image Understanding*, 72(3):404–413.

[13] Sebastian, T. B., Tek, H., Crisco, J. J., Wolfe, S. W., and Kimia, B. B. (1998). Segmentation of carpal bones from 3d ct images using skeletally coupled deformable models. In *MICCAI'98*, pages 1184–1194.

[14] Siddiqi, K., Bouix, S., Tannenbaum, A., and Zucker, S. W. (1999). The hamilton-jacobi skeleton. In *ICCV'99*, pages 828–834, Kerkyra, Greece.

[15] Stetten, G. D. and Pizer, S. M. (1999). Automated identification and measurement of objects via populations of medial primitives, with application to real time 3d echocardiography. In *IPMI'99*, pages 84–97.

[16] Tek, H. and Kimia, B. B. (1997). Volumetric segmentation of medical images by three-dimensional bubbles. *Computer Vision and Image Understanding*, 65(2):246–258.

[17] Wilson, D. L. and Noble, A. (1997). Segmentation of cerebral vessels and aneurysms from mr aniography data. In *IPMI'97*, pages 423–428.

Chapter 12

CLASSICAL AND QUANTUM CONTROLLED LATTICES: SELF-ORGANIZATION, OPTIMIZATION AND BIOMEDICAL APPLICATIONS

Panos M. Pardalos
Center for Applied Optimization, ISE Department,
and Biomedical Engineering Program
University of Florida,
Gainesville, FL 32611-6595, USA
pardalos@cao.ise.ufl.edu

J. Chris Sackellares
Departments of Neurology and Neuroscience,
and Biomedical Engineering Program
Brain Institute and Gainesville V.A. Medical Center,
University of Florida,
Gainesville, FL 3261-6595, USA
sackellares@epilepsy.health.ufl.edu

Vitaliy A. Yatsenko
Scientific Foundation of Researchers and Specialists
on Molecular Cybernetics and Informatics
yatsenko@ise.ufl.edu

Abstract This paper presents new mathematical models of classical (CL) and quantum-mechanical lattices (QML). System–theoretic results on the observability, controllability and minimal realizability theorems are formulated for CL. The cellular dynamaton (CD) based on quantum oscillators is presented. We investigate the conditions when stochastic resonance can occur through the interaction of dynamical neurons with intrinsic deterministic noise and an external periodic control. We found a chaotic motion in phase–space surrounding the separatrix of dynamaton. The suppression of chaos around the hyperbolic saddle arises only

P.M. Pardalos and J. Principe (eds.), Biocomputing, 199-224.

for a critical external control field strength and phase. The possibility of the use of bilinear lattice models for simulating the CA3 region of the hippocampus (a common location of the epileptic focus) is discussed. This model consists of a hexagonal CD of nodes, each describing a controlled neural network model consisting of a group of prototypical excitatory pyramidal cells and a group of prototypical inhibitory interneurons connected via excitatory and inhibitory synapses. A nonlinear phenomenon in this neural network is studied.

Keywords: optimization, control, lattice model, stochastic resonance, chaos, quantum processes, biomedical, self organization, epileptic seizures.

1.   Introduction

In the last century, society was transformed by the conscious application of modern technologies to engines of the industrial revolution. It is easy to predict that in the twenty–first century it will be quantum and biomolecular technologies that influence all our lives. There is a sense in which quantum technology is already having a profound effect [31, 33]. A large part of the national product of the industrialized countries stems from quantum mechanics. We are referring to transistors — the "fundamental particle" of modern electronics. It is now that this information is found to be profoundly different, and in some cases, much more powerful than that based on classical mechanics.

In quantum computing, information is manipulated not discretely, in the classical way, as a series of zeros and ones (bits), but as continuous superposition (qubits) where the number of possibilities is vastly greater. In effect, many computations are performed simultaneously, and calculations that would be intractable classically, become feasible quantally. In this way, the computational theory of — indeed information science itself — is becoming a branch of physics rather than mathematics.

Useful optimization and control of quantum physical processes has been the subject of many investigations [41, 5, 45]. Problems of controlling microprocesses and quantum ensembles were posed and solved for plasma and laser devices, particle accelerators, nuclear power plants, and units of automation and computer technology. Although quantum processors have not yet been developed, there is no doubt that their production is near at hand. Some quantum effects have already been applied in optoelectronic computer devices, but their integration into diverse computer systems still belongs to the future.

Recently there was been an intense interplay between methods of quantum physics, optimization and the theory of control. However the problem of controlling quantum states was in fact brought forth at the rise of quantum mechanics. For instance, many experimental facts of quantum mechanics were established with the use of macroscopic fields acting on quantum ensembles, which from the modern viewpoint can be considered as the control. As the technology of experimentation evolved, new problems in controlling quantum systems arose [41, 5, 45], and their solution required special methods [4, 45].

Optimization problems in quantum and biomolecular computing are also very important. Some exciting results and developments (experimental as well as theoretical) have emerged [16, 37, 38, 44]. Interesting global optimization problems are formulated and developed for processes of nonlinear transformations of information in quantum systems [5, 46] and biomedicine [15, 47]. An important question regarding the optimization of those systems is how to construct efficient controlled mathematical models.

However, many applied optimization problems have yet not been considered. It is necessary to use optimization methods of quantum and biomolecular systems, because of the practical importance of implementation of physical processes satisfying the required quality criterions.

Most of the attention is being focussed on the following problems:

1 Mathematical modeling of controlled quantum and biomolecular systems;

2 Mathematical modeling of controlled physical and chemical processes in the brain. Consider the brain as a quantum macroscopic object [22, 14];

3 Optimal construction of a set of states accessible from a given initial state;

4 Optimization of the set of controls steering the system from a given initial state to a desired accessible state with the greatest or specified probability;

5 Identification of a control that is optimal with respect to a given criterion, e.g. the response time or the minimum number of switches (in bang–bang control);

6 Identification of the measuring operator and the method of its implementation by means of programmed control providing the most reliable measurement at a given instant T;

7 Construction of a system of microscopic feedback providing for the possibility of control with accumulation of data.

In this paper we discuss developments and optimization of classical and quantum-mechanical cellular dynamatons. Cellular dynamata (Fig. 12.1) are complex dynamical systems characterized by two special features: the nodes (component schemes) are all identical copies of a scheme, and they are arranged in a regular spatial lattice.

Definition 1.1 *By a neural dynamical system, (neural dynamaton or ND), we mean a complex dynamical system in which the nodes are all identical copies of a single controlled Hamiltonian dynamical scheme, the standard cell [17].*

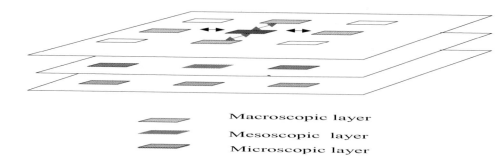

Macroscopic layer
Mesoscopic layer
Microscopic layer

Figure 12.1.   Multilevel neural dynamaton

We consider the cellular dynamaton (CD) [32, 17] in which every cell is considered as the controllable finite-dimensional Hamiltonian system having the form

$$
\begin{aligned}
\partial_t x^k &= \mathcal{F}_H^k(x^k, a^k, u^k), && x^k \in M^k, x^k(0) = x_0^k, \\
\partial_t y_j^k &= -H_j^k(x^k, a^k), && j = 1, \ldots, m, \quad k = 1, \ldots, N \quad (1) \\
u^k &= (u_1^k, \ldots, u_m^k) \in \Omega \subset R^m,
\end{aligned}
$$

where $\mathcal{F}_H^k$ is an integral curve of H^k; $H^k(x^k, a^k, u^k)$ is an arbitrary analytical function of states x^k, of cell interaction parameters a^k and of reconfigurating influences u^k; k is a cell number; $a^k = F^k(x^{k-1}, x^{k+1}, x^k)$. If there exists the Hamiltonian

$$
H^k(x^k, a^k, u^k) = H_0^k(x^k, a^k) - \sum_{j=1}^{m} u_j^k H_j^k(x^k, a^k),
$$

then we have the system of equations

$$
\begin{aligned}
\dot{x}^k &= g_{H_0}^k(x^k, a^k) + \sum_{j=1}^{m} u_j^k g_{H_j}^k(x^k, a^k), \ x^k(0) = x_0^k, \ x^k \in (M^{2n}, \omega), \\
y_j^k &= -\frac{\partial H^k}{\partial u_j^k}(x^k, u^k) \quad j = 1, \ldots, m, \quad k = 1, \ldots, N, \quad (2) \\
u^k &= (u_1^k, \ldots, u_m^k) \in \Omega \subset R^n.
\end{aligned}
$$

Here M is a symplectic manifold with a sympletic form ω and Hamiltonian vector fields $g_{H_j}^k, i = 1, \ldots, m; \Omega \subset R^m$ is a control value domain that contains a point 0; $u^k \in \mathcal{U}^k$ are the control vector components that belong to specified classes $\mathcal{U}^k$ of admissible functions. Consider the control u^k as a generalized external macroscopic influence.

2. Hamiltonian models of the cellular dynamatons

Assume that an elementary cell contained in some CD is described by the Euler–Lagrange equation [29] with external forces; a cell is a nondissipative system with n degrees of freedom $q_1^k, \ldots, q_m^k$ and a Lagrangian $L^k(q_1^k, \ldots, q_m^k, \dot{q}_1^k, \ldots, \dot{q}_m^k)$. Then, the CD dynamics are defined by the equations

$$\frac{d}{dt}\left(\frac{\partial L^k}{\partial \dot{q}^k}\right) - \frac{\partial L^k}{\partial q^k} = F_i^k, \quad i = 1, \ldots, n \tag{3}$$

where $F^k = (F_1^k, \ldots, F_n^k)$ is a vector of generalized external forces. Let some of the vector components be zero and suppose that the rest of them provide the reconfiguration properties possessed by the CD. Here the cell Lagrangian depends explicitly on the vector parameter the current value of which is determined by the neighbouring cell state. Then the representation

$$
\begin{aligned}
\dot{q}_k^k &= \frac{\partial H_0^k}{\partial p_i^k}, & (i = 1, \ldots, n), \\
\dot{p}_i^k &= -\frac{\partial H_0^k}{\partial q_i^k} + u_i^k, & (i = 1, \ldots, m), \\
\dot{p}_i^k &= \frac{\partial H_0^k}{\partial q_i^k}, & (i = 1, \ldots, n),
\end{aligned}
\tag{4}
$$

where $p_i^k = \frac{\partial L^k}{\partial \dot{q}_i^k}$ and H_0^k is an internal Hamiltonian for the k-th cell, is true.

We have the system

$$\dot{x}^k = g_{H_0}^k(x^k) + \sum_{j=1}^{m} u_j^k g_{H_j}^k(x^k), \quad x^k(0) = x_0^k, \ x^k \in M^k, \tag{5}$$

at the arbitrary coordinates x^k for a smooth manifold M^k; M^k is a symplectic manifold with a symplectic form ω^k; $g_H^k(x^k)$ and $(j = 0, 1, \ldots, m)$ are Hamiltonian vector fields satisfying the relation $\omega^k(\mathcal{F}_{H_j}) = -dH_j^K$. According to Darboux's theorem [2, 1], there exist canonical coordinates (p^k, q^k) such that

$$\omega^k = \sum_{i=1}^{m} dp_i^k \wedge dq_i^k. \tag{6}$$

from the point of view of locality.

If it is assumed that an interaction Hamiltonian H_j^k is equal to q_j ($j = 1, \ldots, m$). Then the system of equations (3) is derived. A CD cell 'state–output' map may be represented by the expression

$$y_j^k = H_j^k(x^k) \quad (j = 1, \ldots, m), \tag{7}$$

i.e., an output signal is specified by a disturbance $u_1^k F_H^k, \ldots, u_m^k F_H^k$ for a natural input, and this circumstance, in its turn, is the reason why the energy $H_1^k, \ldots, H_m^k$ values are changed.

Introduce the following definitions.

Definition 2.1 *A controlled CD model defined by*

$$\begin{aligned}
\partial_t x^k &= G^k(x^k, u^k) \quad x^k(0) = x_0^k \quad x \in M^k, \\
y_j^k &= h_j^k(x^k, u^k) \quad j = 1, \ldots, r \quad u^k = (u_1^k, \ldots, u_m^k) \in \Omega \subset R^n \tag{8}
\end{aligned}$$

is called the Hamiltonian model if it admits the representation in the form of the system

$$\partial_t x^k = \mathcal{F}_{H^k}^k(x^k, u^k), \quad x^k \in M^k, \ x^k(0) = x_0^k, \tag{9}$$

where $H^k(x^k, u^k)$ is the arbitrary analytical function of the k-th cell state x^k and of the control parameters u^k; $\mathcal{F}_{H^k}^k$ is integral curve of H^k. If $H^k(x^k, u^k)$ be of the form

$$H_0^k(x^k) - \sum_{j=1}^m u_j^k H_j^k(x^k), \tag{10}$$

then system (1) is obtained. CD cell outputs are specified here by the expression

$$y_j^k = -\frac{\partial H^k}{\partial u_j^k}(x^k, u^k) \quad (j = 1, \ldots, m). \tag{11}$$

Since the state space M^k for system (5) is symplectic, it is possible to yield the Poisson bracket

$$\{F^k, G^k\} = \sum_{j=1}^n \left(\frac{\partial F^k}{\partial p_i^k} \frac{\partial G^k}{\partial q_i^k} - \frac{\partial F^k}{\partial q_i^k} \frac{\partial G^k}{\partial q_i^k} \right), \tag{12}$$

$$F^k, G^k : M^k \to R.$$

Definition 2.2 *Let L^k be the Lie algebra for the Hamiltonian vector fields of some k-th cell. The linear span for the functions $f^r(H_j^k)$, where $f^k \in L^k$, is termed the k-cell observation space $\mathcal{H}^k$.*

Since $X_F^k(G^k) = \{F^k, G^k\}$ *and* $[g_F^k, g_G^k] = g_{\{F,G\}}^k$, $\mathcal{H}_j^k$ *is defined by the functions*

$$\{F_1^k, \{F_2^k, \{\ldots\{F_k^k, H_k^k\}\ldots\}\}\},\qquad\qquad(13)$$

for CD (5), (7) and here F_r^k, $r = 1, \ldots, k$ *is equal to* H_i^k, $i = 0, 1, \ldots, m$.

If the theorems considered for the case with nonlinear system controllability and observability [4] are applied for a CD, then the following results are obtained.

Proposition 2.1 $\mathcal{H}^k$ *is an ideal formed by* $H_1^k, \ldots, H_m^k$ *in the Lie algebra (under Poisson bracket) generated, in its turn, by* $H_0^k, \ldots, H_m^k$.

Proposition 2.2 *Let (5), (7) be a Hamiltonian CD model. Then:*

a) *The CD states are strongly accessible and the CD is weakly observable if* $\dim d\mathcal{H}^k(x^k) = \dim M^k, \forall x^k \in M^k$; *otherwise the CD is quasiminimal.*

b) *The CD is strongly accessible and observable if* $\dim d\mathcal{H}^k(x^k) = \dim M^k, \forall x^k \in M^k$ *and* $\mathcal{H}^k$ *allows to distinguish* M^k *points; otherwise the CD is minimal.*

A nonminimal CD can be reduced to a quasiminimal type under the same 'input-output' map. If a CD is Hamiltonian, then the Hamiltonian system is yielded again as the result of the transformation procedure.

It was mentioned earlier that the quasiminimal CDs with one and the same 'input-output' map are locally diffeomorphic. In the Hamiltonian case, map equivalence means the smooth simplectomorphism, i.e. there are the local canonical transformation and the energy equivalence within the constants (for both CDs). Therefore, the 'input-output' map determines not only the CD state, but also the canonical CD state structure and the internal energy. This approach can be extended to general-type CDs (1) and here the observation space $\mathcal{H}$ is used, i.e., $\mathcal{H}$ is the linear space that exists for the functions (x, u) and includes $\partial H_1/\partial u, \ldots, \partial H_m/\partial u$ and invariants. A Lie algebra forms by calculations of Poisson brackets of Hamiltonians, Hamiltonian derivatives, etc.

When the well-known accessibility and observability theorems mentioned in [4, 23] are considered for the CDs cases, they are formulated as follows.

Theorem 1 *Let (5,7) be strongly accessible and observable by a CD. Then a CD is minimal if and only if the vector field Lie algebra is self-adjoint.*

Proof: The proof of the theorem follows from the condition of minimality of Hamilton systems [43], and the necessary and sufficient conditions of controllability for Lie–determined systems [4, 23].

Theorem 2 *A strongly accessible and observable CD is self-adjoint (and, therefore, Hamiltonian), if the cell output variations* $\delta_1 y^k$, $\delta_2 y^k$ *are determined on a compact [0,T] and the equality*

$$\int_0^\infty [\delta_1^T u^k(t)\delta_2 y^k(t) - \delta_2^T u^k(t)\delta_1 y^k(t)] = 0.$$

is true for an arbitrary piecewise constant u_j^k *and for any two cell input variations* $\delta_1 u^k$, $\delta_2 u^k$.

Proof: This result is a synthesis of results from the theory of controlled Hamilton systems put into focus through the maximum principle [23, 43].

Besides controllability and observability duality, the Hamiltonian CAs possess a number of other useful features. In particular, the Hamiltonian CD theory is sometimes essential for the quantum models of adaptive computational medium (ACM). The global optimization algorithms [21] seem to be solve the problem of optimal nonlinear information transforming in ACM.

3.   Self-organization of the neural networks

The phenomenon of self–organization [32] that arises due to the interplay of noise and an external control in a bistable elementary cell (EC) has attracted considerable attention in recent years. In a bistable cell characterized by a double–well potential, noise may induce transitions between the two stable states of the system, which behaves stochastically and consequently has a continuous power spectrum. The application of an external time–periodic control has been shown to produce a sharp enhancement of the signal power spectrum about the forcing frequency. This effect has been observed experimentally in simple bistable electronic and optical CD and in more complex systems. Based on this idea, new techniques for extracting periodic signals from the noise background have been used [28]. The noise commonly encountered in this context is of external origin in the sense that it appears as a separate term in the appropriate Langeven equations. However, noise or stochasticity also arises in the deterministic dynamical evolution of a bistable cell when coupled to another autonomous or nonautonomous one-degree-of-freedom system.

It is interesting to know the movement property of these systems around the separatrix [32]. A characteristic motion of these systems, important in this context, is the motion around the separatrix. It is well known that a generic Hamiltonian perturbation always yields chaotic motion in a phase–space layer surrounding the separatrix. This chaos appearing in the vicinity of the hyperbolic saddle of the bistable potential, although remaining confined within Kolmogorov – Arnold – Moser (KAM) barriers [2] (which separate the various resonance zones) and being local in character, serves as a precursor to global deterministic stochasticity. The object of the present section is to examine whether something like stochastic resonance can occur through the interaction of a dynamical bistable EC with intrinsic deterministic noise of this kind and an external periodic forcing. We show that deterministic stochasticity or chaos can be inhibited by critically adjusting the phase and amplitude of the applied resonant driving field.

Our analysis is based on a simple coupled oscillator model describing a nonlinear oscillator with a symmetric double-well potential quadratically coupled to a harmonic oscillator. This dynamical system admits of chaotic behavior and by virtue of having a homoclinic orbit in addition to periodic orbits is amenable to theoretical analysis using Melnikov's technique.

In Melnikov's method one is concerned with the perturbation of the homoclinic manifold in a Hamiltonian system which consists of an integrable part and a small perturbation. It is well known that if Melnikov's function, which measures the leading nontrivial distance between the stable and the unstable manifolds, allows simple zero, then the stable and the unstable manifolds, which for an unperturbed system of oscillators coincide as a smooth homoclinic manifold, intersect transversely for small perturbation generating scattered homoclinic points. This asserts the existence of a Smale's horseshoe [34] on a Poincare map and qualitatively explains the onset of deterministic stochasticity around the separatrix. We show that, if an additional external resonant periodic control is brought into play, then, depending on its amplitude and phase, the Melnikov function can be prevented from admitting simple zeros, which implies that resonance restores the regularity in the dynamics. In other words, the deterministic stochasticity is inhibited.

To start with, we consider the following theorem.

Theorem 3 *Assume the Hamiltonian dynamaton (HD) is given by*

$$H^k(q^k, p^k, x^k, v^k) = G^k(x^k, v^k) + F^k(q^k, p^k) + \epsilon H^k_{(1)}(q^k, p^k, x^k, v^k), \quad (14)$$

where

$$G^k(x^k, v^k) = \frac{1}{2}\{(v^k)^2 + \omega^2(x^k)^2\}, \quad (15)$$

$$F^k(q^k, p^k) = \frac{1}{2}(p^k)^2 - \frac{1}{2}(q^k)^2 + \frac{1}{4}(q^k)^4, \quad (16)$$

and

$$H^k_{(1)}(q^k, p^k, x^k, v^k) = \frac{\varrho}{2}(x^k - q^k)^2 \quad (17)$$

denote the harmonic oscillator, the nonlinear oscillator with a bistable potential, and the coupling perturbation, respectively, ϱ and ϵ are the coupling and smallness parameters; $I^k = \{1, \ldots, N\}$; N is the number of nodes, equally spaced in the physical space; $\epsilon = \zeta(q^{k+1}, p^{k+1}, x^{k+1}, v^{k+1}, q^{k-1}, p^{k-1}, x^{k-1}, v^{k-1})$ is local bias.

1 *If HD consists of F^k and G^k subsystems, then HD simulates an homoclinic and periodic orbits;*

2 *If HD consists of F^k, G^k and $H^k_{(1)}$ subsystems, and sufficiently small $\epsilon > 0$, then HD simulates the dynamical chaos around the hyperbolic saddle of the bistable potential of each cell.*

3 *If HD consists of $F^k, G^k, H^k_{(1)}$,*

$$H^k_{(2)} = Aq^k \cos(\Omega t + \phi),$$

$\epsilon > 0$ is sufficiently small, $A = \varrho(2\bar{h})^{1/2}/\Omega$, $\Omega = \omega$, and $\phi = -\pi/2$

then HD produces a stochastic resonance.

Proof: The canonically conjugate pairs of coordinates and momenta for G^k and F^k systems are (x^k, v^k) and (q^k, p^k), respectively. ω is the angular frequency of the harmonic oscillator. The uncoupled system consisting of F^k and G^k systems is integrable. The Hamiltonian perturbation $H^k_{(1)}$ breaks the integrability by introducing horseshoes into the dynamics and thereby making the system chaotic.

Making use of a canonical change of coordinates to action-angle (I^k, θ^k) variables, where θ^k is 2π periodic and $I^k \geq 0$, one obtains G^k as a function of I^k alone as follows:

$$G^k = \omega I^k. \tag{18}$$

The action and angle variables for G^k are expressed through the relations

$$\begin{aligned}
x^k &= (2I^k/\omega)^{1/2} \sin \theta^k, \\
v^k &= \omega(2I^k/\omega)^{1/2} \cos \theta^k.
\end{aligned} \tag{19}$$

The integrable equations of motion are $(\epsilon = 0)$

$$\dot{q}^k = \frac{\partial F^k}{\partial p^k}, \quad \dot{p}^k = -\frac{\partial F^k}{\partial q^k}, \tag{20}$$

$$\dot{\theta}^k = \omega, \quad \dot{I}^k = 0.$$

The Hamiltonian system associated with the F^k system possesses the homoclinic orbit

$$\begin{aligned}
q^k(t) &= (2)^{1/2} \operatorname{sech}(t - t_0), \\
p^k(t) &= -(2)^{1/2} \operatorname{sech}(t - t_0) \tanh(t - t_0),
\end{aligned} \tag{21}$$

joining the hyperbolic saddle ($q^k = 0$, $p^k = 0$) to itself. The G^k system (18) contains 2π periodic orbits

$$\theta^k = \theta_0 + \omega t,$$
$$I^k = I_0^k,$$
(22)

where θ_0^k and I_0^k are determined by the initial conditions. Thus for the uncoupled system $F^k \times G^k$ we have the products of homoclinic and periodic orbits.

Let us now turn toward the perturbed Hamiltonian H^k. The perturbation $H_{(1)}^k$ is smooth. Also, the total Hamiltonian H^k is an integral of motion. The equations of motion are

$$\dot{q}^k = \frac{\partial F^k}{\partial p^k} + \epsilon \frac{\partial H_{(1)}^k}{\partial p^k}, \quad \dot{p}^k = -\frac{\partial F^k}{\partial q^k} - \epsilon \frac{\partial H_{(1)}^k}{\partial q^k},$$

$$\dot{\theta}^k = \omega + \epsilon \frac{\partial H_{(1)}^k}{\partial I^k}, \quad \dot{I}^k = -\epsilon \frac{\partial H_{(1)}^k}{\partial \theta^k}.$$
(23)

For $\epsilon > 0$, but small, one can show that transverse intersection occurs.

Following [32] the two-degree-of-freedom autonomous system (23) can be reduced to a one-degree-of-freedom nonautonomous system using the classical reduction method. In the process one eliminates the action I^k from equation (23) using the integral of motion H^k (14). One then further eliminates the time variable t, which is conjugate to H^k, and the resultant equations of motion are written by expressing the coordinate and momentum as functions of the angle variable θ^k.

One needs not follow here this procedure explicitly, but can directly use the theorem to calculate the Melnikov's function [35], which measures the leading nontrivial distance between the stable and unstable manifolds in a direction transverse to dynamic variable θ^k. In practice, the calculation involves the integration of Poisson bracket $\{F^k, H_{(1)}^k\}$ around the homoclinic orbit as follows:

$$\Upsilon(t_0) = \int_{-\infty}^{+\infty} \{F^k, H_{(1)}^k\} \, dt, \quad k = 1, \ldots, N$$
(24)

Explicit calculation of the Poisson bracket using equation (16),(17), and (21) yields

$$\{F^k, H_{(1)}^k\} = \frac{\partial F^k}{\partial q^k} \frac{\partial H_{(1)}^k}{\partial p^k} - \frac{\partial F^k}{\partial p^k} \frac{\partial H_{(1)}^k}{\partial q^k} = 2\varrho \operatorname{sech}^2(t - t_0) \tanh(t - t_0) +$$
$$+ \quad 2\varrho (I^k/\omega)^{1/2} \sin \omega t \operatorname{sech}(t - t_0) \tanh(t - t_0).$$
(25)

The Melnikov function is then given by

$$\Upsilon(t_0) = \frac{2\varrho(h)^{1/2}}{\omega} \cos \omega t_0 \left| \pi\omega \operatorname{sech} \frac{\pi\omega}{2} \right|. \tag{26}$$

In calculating the relation (26) one must take into account that the energy of the homoclinic orbit is zero and $I^k = (h-0)/\omega, h > 0$, where

$$H^k(q^k, p^k, x^k, v^k) = h.$$

Since $\Upsilon(t_0)$ has simple zeros and is independent of ϵ, we conclude that for $\epsilon > 0$, but sufficiently small, one can have transverse intersection (and horseshoes on the Poincare map) on the energy surface $h > 0$. What follows immediately is that we have simulated the dynamical chaos around the hyperbolic saddle of the bistable potential through the transverse intersection, which is probed by Melnikov's function.

Let us now see the effect of an external time-dependent periodic driving force on this chaos. For this we introduce the perturbation $H^k_{(2)}$, which is of the same order $[O(\epsilon)]$ as $H^k_{(1)}$,

$$H^k_{(2)} = Aq^k \cos(\Omega t + \phi), \tag{27}$$

where A, Ω, and ϕ denote the amplitude, the frequency, and the phase of the external field.

It is immediately apparent that the energy function

$$H^k = F^k(q^k, p^k) + G^k(I^k) + \epsilon H^k_{(1)} + \epsilon H^k_{(2)} \tag{28}$$

is no longer conserved and one has to consider an equation [35] for the time development of H^k in addition to Hamilton's equations of motion

$$\dot{q}^k = \frac{\partial F^k}{\partial p^k} + \epsilon \frac{\partial H^k_{(1)}}{\partial p^k}, \quad \dot{p}^k = -\frac{\partial F^k}{\partial q^k} - \epsilon \frac{\partial H^k_{(1)}}{\partial q^k}, \tag{29}$$

$$\dot{\theta} = \omega + \epsilon \frac{\partial H^k_{(1)}}{\partial I^k}, \quad \dot{I}^k = -\epsilon \frac{\partial H^k_{(1)}}{\partial \theta},$$

where

$$f^k = A \cos(\Omega t + \phi). \tag{30}$$

One can then use again the classical reduction scheme [35] along with an average $\bar{h}$ instead of h (although there are several averaging procedures [35] we do not need to have an explicit expression for $\bar{h}$ for our purpose). The relevant Melnikov's function for the problem is as follows:

$$\Upsilon^k_f(t_0) = (1/\omega^2) \left[\Upsilon(t_0) + \int_{-\infty}^{+\infty} \left(-\frac{\partial F^k}{\partial p^k} f^k \right)_{t-t_0} dt \right], \tag{31}$$

where $\Upsilon(t_0)$ is given by (26) with the replacement of h by some average $\bar{h}$ appropriate for the time-dependent H^k (28). The integrand in the expression is a function of the time interval $t - t_0$, where t_0 refers to the time of intersection. On explicit calculation of the integral in (31), making use of $\partial F^k/\partial p^k = p^k$ and equation (21) we obtain

$$\int_{-\infty}^{+\infty} (\cdots) \, dt = (2)^{1/2} A \sin(\Omega t_0 + \phi) \left[\pi \Omega \operatorname{sech} \frac{\pi \Omega}{2} \right]. \qquad (32)$$

Also, we have

$$\Upsilon(t_0) = \frac{2\varrho(\bar{h})^{1/2}}{\omega} \cos \left[\pi \Omega \operatorname{sech} \frac{\pi \omega}{2} \right]. \qquad (33)$$

Let us consider an interesting situation. For $\phi = -\pi/2$ and $\Omega = \omega$, a resonance condition, if one chooses the amplitude of the driving field as

$$A = \varrho(2\bar{h})^{1/2}/\Omega, \qquad (34)$$

then the Melnikov's function $\Upsilon_f^k(t_0)$ vanishes, i.e., ceases to have simple zeros. This implies that the resonance inhibits transverse intersections of the stable and unstable manifolds. As a result, the regularity is restored in the EC and we have a typical situation similar to what is called the stochastic resonance. For $A \neq \varrho(2\bar{h})^{1/2}$ the Melnikov's function, however, has simple zeros and the dynamics is chaotic. It has to be noted further that in the absence of any one of the perturbation terms the transverse intersections occur and the system becomes chaotic. It is the crucial interplay of both these terms as expressed through the condition (34) that leads to the inhibition of chaos.

It is pertinent to note the important distinction between stochastic resonance and the suppression of chaos as considered in the present paper. The stochastic resonance mechanism is related to the oscillating behavior of the signal autocorrelation function for times larger than the relevant decay time (reciprocal of the Kramers rate) in the unperturbed bistable system, and such an effect is apparent even when the perturbation is weak enough not to appreciably affect the rate of noise-induced switch process [28]. The suppression of chaos around the hyperbolic saddle, on the other hand, arises only for a critical external field strength and phase. The latter has apparently no role in the stochastic resonance. Last we mention that, although we have considered the motion around the separatrix corresponding to a bistable potential, we hope that such an inhibition effect can be seen for other types of potentials where the separatrices do exist.

4.   Bilinear lattices and epileptic seizures

The dynamics of synapses and global optimization. A simple version of the dynamics of neural synapses can be written in the form [24]

$$\dot{S}_{kj} = \alpha i_k f_j(t) - H, \tag{35}$$

where H defines decay; i is the electric current; $S_{kj} = V_{ion}s_{kj}$; V_{ion} is the chemical potential; α is the learning rate; s is the action potential; k, j are cell numbers; $f_j(t)$ is the instantaneous firing rate of neuron j. Decay terms, involving i_k and $f_j(t)$, are essential to forget old information. The learning rate α might also be varied by neuromodulator molecules which control the overall learning process. In the case $S_{kj} = S_{jk}$, there is a Lyapunov or "energy" function for equation (35)

$$E = -\frac{1}{2}\sum S_{ij}V_iV_j - \sum I_iV_i + \frac{1}{\tau}\sum \int V^{-1}(f')df', \tag{36}$$

and the quantity f_i always decreases in time [20]. The dynamics then are described by a flow to an attractor where the motion ceases. The existence of this energy function provides a programming tool [19]. In the high–gain limit the system has direct relationship to physics. It can be started most simply when the asymptotic values are scaled to ± 1. The stable points of the dynamic system then have each $V_i = \pm 1$, and the stable states of the dynamic system are the stable points of an Ising magnet with exchange parameters $J_{ij} = S_{ij}$. Many difficult computational problems can be posed as global optimization problems [38]. If the quantity to be optimized can be mapped onto the form equation (35), it defines the connections and the "program" to solve the global optimization problem.

Bilinear lattice model. The bilinear lattice model based on a two–variable reduction of the Hodgkin–Hukley model [18] was initially proposed by Morris–Lecar [36] as a model for barnacle muscle fiber, but is of general utility in modeling the pyramidal cells in network of the CA3 region of the hyppocampus (Fig. 12.2).

The system of equations for our proposed network model is

$$\dot{x}_1^i = a_1 + b_{11}x_1^i + c_{12}x_2^i u_1^i + u_2^i + b_{13}x_3^i, \tag{37}$$

$$\dot{x}_2^i = a_2 + b_{22}x_2^i, \tag{38}$$

$$\dot{x}_3^i = b_{31}x_1^i + d_3 u_2^i, \tag{39}$$

$$y^i(t) = Ex^i(t). \tag{40}$$

This system can by represented by a pair of equations of the form

$$\dot{x}^i(t) = A + \left(B + u_1^i(t)C\right)x^i(t) + Du_2^i, \tag{41}$$

$$y^i(t) = Ex^i(t), \tag{42}$$

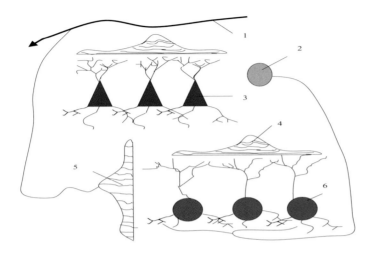

Figure 12.2. A network showing interconnections between an excitatory pathway, a population of pyramidal cells and a population of inhibitory interneurons (1 is excitatory pathway, 2 is population of excitatory synapses, 3 is pyramidal neurons, 4, 5 are populations of inhibitory synapses, and 6 depicts inhibitory neurons)

with the A, B, C, D, E matrices and the u_j is a scalar function of time and x^i_j. Here

$$A = \begin{bmatrix} a_1 \\ a_2 \\ 0 \end{bmatrix}, B = \begin{bmatrix} b_{11} & 0 & b_{13} \\ 0 & b_{22} & 0 \\ b_{31} & 0 & 0 \end{bmatrix}, C = \begin{bmatrix} 0 & c_{12} & 0 \\ 0 & 0 & 0 \\ 0 & 0 & 0 \end{bmatrix}, D = \begin{bmatrix} 1 \\ 0 \\ d_3 \end{bmatrix},$$

$a_1 = g_{ca}m_\infty + g_L V^L; a_2 = (\phi w_\infty)/\tau_w; b_{11} = -g_{ca}m_\infty - g_L; b_{13} = -\alpha_{inh}; b_{22} = -\phi/\tau_w; b_{31} = \alpha_{exc}; c_{12} = -g_k; d_3 = bc; m_\infty = f_1(x^i_1, y_1); w_\infty = f_2(x^i_1, y_3, y_4); \alpha_{exc} = f_3(x^i_1, y_5, y_6); \alpha_{inh} = f_4(x^i_1, y_6, y_7); \tau_w = f_5(x^i_1, y_3, y_4); u^k_1 = (x^k_1 - V^K_k); u^k_2 = I^k; x^k_1$ and x^k_3 are the membrane potentials of the pyramidal and inhibitory cells, respectively; x^k_2 is the relaxation factor wich is essentially the fraction of open potassium channels in the population pyramidal cells; all three variables apply to node k in the lattice. The parameters g_{ca}, g_k and g_L are the total conductances for the populations of Ca, K and leakage channels, respectively. V^K_k is the Nerst potential for potassium in node. The parameter V_L is a leak potential, τ_w is a voltage dependent time constant for W_i, I^k is the applied current, and ϕ and b are temperature scaling factors. The parameter c differentially modifies the current input to the inhibitory interneuron. The parameter V^K_k, the equilibrium potencial of potassium for node "k", is taken to be a function of the average extracellular potassium of the six nearest neighbors "k". The value if V_k^K is changed after each interval of simulation time using the following equation:

$$V_k^K = \left(\sum_{j=1}^{6} \frac{V_j^0}{6} \right) - \frac{1}{2},$$

where

$$V_j^0 = \frac{1}{T} \sum_{t_{int}/6}^{t_{int}} x_j^k(t)\,dt;$$

t_{int} determines how long the node's dynamics will remain autonomous without communication with other nearest neighbor nodes, and $T = t_{int} - t_{int}/6$ is the interval of time in computing the averages.

This model has a rich variety of relevant dynamical behaviors. The bilinear lattice model can reproduce single action potentials as well as sustained limit cycle oscillations for different values of the parameters. Simulation with the proposed bilinear model have shown simple limit cycles as well as periodic state and aperiodic behavior. The periodic state corresponds to a phase–locked mixed mode state on a torus attractor. As the parameter c is controlled we see a large variety of a mixed mode states interspersed with regions of apparent chaotic behaviour [30]. The dynamics are less complex for smaller values of c, i.e., when the current to the inhibitory cells is relatively low compared to the current input to the pyramidal cells. A low degree of inhibition results in system that is more likely to be periodic, suggesting the type of spatiotemporal coherence that could exist with a seizure. When inhibition is completely absent, the dynamics of a system goes to a fixed point corresponding to a state of total depolarization of the network. An intersection of the transition regions between mixed–mode states undergo a periodic–doubling sequence as the underlying torus attractor breaks up into a fractal object. When we control the parameters b and y_6 similar behavior is seen in a series of system investigations.

Chaotic model of lattice. Bilinear point mappings are self-maintained and a comparatively new part of the theory of dynamic systems, where objects with continuous and discrete time are studied. The following mapping is the particular case of this class [32]

$$x_{n+1} = \sum_{i=1}^{m} a_i u_i(n)(1 - x_n), \tag{43}$$

where a_i is a constant parameter; u_i is a scalar control, which can be chosen in a feedback form.

For $a_\infty < a \leq 4$, where $a_\infty = 3.5696\ldots$ the equation has cycles with arbitrary periods including aperiodic trajectories. For $a = 4$ the system dynamics

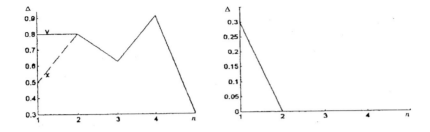

Figure 12.3. Synchronism regime

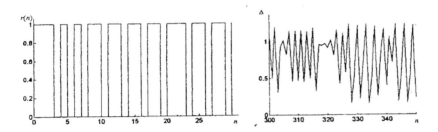

Figure 12.4. "Almost synchronism" regime

has the property of ergodicity and mixing with exponential divergence of close trajectories.

For (43) mappings and

$$y_{n+1} = bu(n)(1 - x_n) \tag{44}$$

it is possible to construct a formal mathematical model of the neuron

$$\begin{pmatrix} x_i(n+1) \\ y_i(n+1) \end{pmatrix} = \frac{1}{1 + 2D_i(n)} \times \begin{pmatrix} 1 + D_i(n) & D_i(n) \\ D_i(n) & 1 + D_i(n) \end{pmatrix} \begin{pmatrix} f[x_i(n)] \\ g[y_i(n)] \end{pmatrix}. \tag{45}$$

Here D is the interrelation coefficient between systems (43) and (44), $f(x) = ax(1 - x)$, $g(y) = by(1 - y)$, i is an order number of the neuron.

For $a = b$ and for large D (the synchronism regime) we get the solution represented in Fig. 12.3.

For $a \approx b$ ("almost synchronism") the possibility of realization of the regime, which is characteristic for neuron dynamics under the conditions

$$\Delta = |x_n - y_n|, \tag{46}$$

$$r(n) = \begin{cases} 1, & \text{for} \Delta \le \varepsilon \\ 0, & \text{for} \Delta > \varepsilon \end{cases}, \tag{47}$$

where ε is a small parameter.

Fig. 12.4 shows the dependence of Δ and r with respect to n ($D = 10^4, a = 3.9001, b = 3.9$).

The first 300 points were omitted on order to obtain a stable process. For construction $r(n)$ 30 points of a steady-state process were used.

5.   Quantum model of neural networks

We proceed from the conventional H–Hamiltonian quantization scheme according to which generalized coordinates $q_1, \ldots, q_n$ and momenta $p_1, \ldots, p_n$ are quantized first of all and then H is expressed with respect to these quantum values. Pursuant to the concepts stated in [8], when a function set on M is quantized, such quantization means that the Hermitian operator f that acts on a complex Hilbert space is brought into correspondence with each function contained in this set. Besides this, such operators must satisfy some commutative relations which, in their turn, correspond to Poisson bracket of the classical functions. Assume $M = R^{2n}$, and let $q_1, \ldots, q_n$ and $p_1, \ldots, p_n$ be the variables that must be quantized. Then the following commutation relations are true:

$$[\hat{q}_i, \hat{q}_j] = 0, \quad [\hat{p}_i, \hat{p}_j] = 0, \quad [\hat{q}_i, \hat{p}_j] = i\hbar\delta_{ij}\hat{I}. \tag{48}$$

Let us reduce the value $\hbar$ to 1 in order to simplify our further discussion. Since a skew-Hermitian operator $-i\hat{f}$ corresponds to every Hermitian operator $\hat{f}$, then relations (48) mean that the map $\hat{f} \rightarrow -i\hat{f}$ is the morphism of Lie algebra (which contains the Poisson bracket and which is built up on the basis of the classical functions $(q_1, \ldots, q_n, p_1, \ldots, p_n)$ into the skew-Hermitian operator Lie subalgebra. Here the Poisson bracket [f,g] is mapped into the bracket $-i[\hat{f}, \hat{g}] = i[-i\hat{f}, -i\hat{g}]$.

The traditional scheme according to which the values $q_1, \ldots, q_n$ and $p_1, \ldots, p_n$ are quantized and which satisfies expressions (48) means that the Hilbert space $L^2(R^n, C)$ is introduced, that the operator $\hat{q}_j$ is assigned to a coordinate q_j (multiplication by q_j) and that the operator $\hat{p}_j = i\frac{\partial}{\partial q_j}$ is assigned to p. The same scheme may also be applied for some simple Hamiltonian CAs.

Example. Consider the harmonic oscillator possessing an external force u^k which, in its turn, may depend on quantum-mechanical observables. The Hamiltonian for such system is specified by the expression

$$H^k(q^k, p^k, u^k) = (p^k)^2/2m + 1/2a^k(q^\ell)(q^k)^2 - u^k(t)q^k. \tag{49}$$

The quantization results in the relation

$$i\hbar\frac{\partial\Psi^k}{\partial t} = -\frac{1}{2m}\frac{\partial^2\Psi^k}{\partial(q^k)^2} + \left[\frac{1}{2}v^k(t)a^k(q^\ell)(q^k)^2 - u^k(t)q^k\right]\Psi^k \tag{50}$$

$$\Psi^k \in L^2(R, C),$$

which describes, in particular, the dynamics of a particle contained in a single well (bipotential) and this particle is affected, in its turn, by a homogeneous classical external field, and here a field value and a field direction are the arbitrary time functions. Generally, when the Hamiltonians $H_0, \ldots, H_m$ are transformed to the Hermitian operators present in some complex Hilbert space $\mathcal{H}$, then the quantized system looks like

$$i\frac{\partial \Psi^k}{\partial t} = \left(\hat{H}_0^k - \sum_{j=1}^{m} u_j^k \hat{H}^k \right) \Psi^k, \quad \Psi^k \in \mathcal{H} \tag{51}$$

(Schrödinger representation). Here $\hat{H}^0$ is the Hamiltonian for an isolated quantum system. The second term contained in this expression, represents the interaction with the external sources through the system of the interaction Hamiltonians $\hat{H}_1^k, \ldots, \hat{H}_m^k$.

Note that if there is a complex Hilbert space $\mathcal{H}$, then the imaginary part of the inner Hermitian product $\langle\,\rangle$ defines a simplectic form on $\mathcal{H}$. With respect to this simplectic form, the skew-adjoint operators $-i\hat{H}k_j$ are the linear Hamiltonian vectorfields on $\mathcal{H}$. Such fields are the expected observables H_j^k, i.e., they correspond to the quadratic Hamiltonians

$$\langle H_j^k \Psi^k, \Psi^k \rangle = \langle \Psi^k \mid H_j^k \mid \Psi^k \rangle, \quad j = 0, \ldots, m. \tag{52}$$

That is why expression (50) provides the Hamiltonian system

$$\frac{\partial \Psi^k}{\partial t} = -i\hat{H}_0^k \Psi^k + \sum_{j=1}^{m} u_j(-i\hat{H}_j^k)\Psi^k \tag{53}$$

$$y_j^k = \langle \Psi^k \mid H_j^k \mid \Psi^k \rangle, \quad j = 1, \ldots, m \tag{54}$$

with the macroscopic controls $u_j^k, \ldots, u_m^k$ and outputs $y_1^l, \ldots, y_m^k$. The outputs on the infinite-dimensional state space $\mathcal{H}$ are equal to the expected observable H_j^k (Dirac, 1958). Note that the essence of the fact that y_j^k is equal to the expected values of $\hat{H}_j^k$ does not arise from the measurement problem. In addition, although $\hat{H}_j^k$ measurement process introduces disturbances into a system, these disturbances propagate along the channels that correspond, in their turn, to the inputs $\hat{u}_j^k$. Since H^k is the simplectic space, then Poisson bracket for two observable values, namely, $\hat{H}_1^k$ and $\hat{H}_2^k$, is calculated in the following way

$$\langle \hat{\Psi}^k \mid \hat{H}_1^k \mid \Psi^k \rangle, \quad \langle \Psi^k \mid \hat{H}_2^k \mid \Psi^k \rangle = -\langle \Psi^k \mid [\hat{H}_1^k, \hat{H}_2^k] \mid \Psi^k \rangle. \tag{55}$$

The sign "−"arises if the commutators existing for $-i\hat{H}_1^k$ and $-i\hat{H}_2^k$ are taken into account. Therefore, the observation space for a quantum-mechanical CD is specified by the ideal generated by the expected values of $\hat{H}_1^k, \ldots, \hat{H}_m^k$ within the Lie algebra generated by the expected values of $\hat{H}_0^k, \ldots, \hat{H}_m^k$ under the Poisson bracket. Let us compare this observation space with the observation space existing for system (1). If the quadratic-linear Hamiltonian are present, then according to Erenfest theorem, these two observation spaces are equal (in the sense that there are the averaged values which satisfy the classical equations). Therefore, the transputer quantum-mechanical CD can never be minimal. Finally, note that CD quantization may be started not with the variables $q_1^k, \ldots, q_n^k$ and $p_1^k, \ldots, p_n^k$, but with the observation space existing for system (9), (11). If there are the CDs based on harmonic oscillators, the essense of the matter is not changed because the observation space is determined by the linear span of q^k, p^k and 1. When it is necessary to quantize the observation spaces existing for the CDs of different types, then such approach is of interest.

Finally, it is actually possible to quantize the externally and nonpotentially reconfigurable Hamiltonian CAs. But the present paper states that it is possible to quantize the "completed" transputer-type CAs.

6. Concluding remarks

This paper proposes Hamiltonian models of classical and quantum–mechanical networks. The observability, controllability and minimal realizability theorems for the neural networks are formulated. The cellular dynamaton based on quantum oscillators is presented. The proposed models are useful to investigate information processes in the human brain. Quantum models could make it possible to view the brain as a quantum macroscopic object with new physical properties [22]. These relation properties are very much needed in neuroscience for simulation of physical properties in the brain [14]. However, efficient techniques to solve reconstruction problems of dynamical systems and optimization algorithms are needed. The global optimization algorithms, described in [21], can be used.

Self–organization and stochastic resonance can occur through the interaction between macroscopic and microscopic levels of neural dynamaton with intrinsic deterministic noise and an external control forcing. These interactions through mesoscipic dynamics [13] are the basis for self–organization of brain and behaviour. The neural network controlled by noise may induce transitions between the stable states of the neurons, which behaves stochastically and consequently has a continuous power spectrum. The suppression of chaos around the hyperbolic saddle arises only for a critical external field strength and phase. The latter has apparently no role in the stochastic resonance.

The controlled cellular dynamaton can be used for modeling the CA3 region of the hippocampus (a common location of the epileptic focus). This region is the self–organized information flow networks of the human brain. This model consists of a hexagonal CD of nodes, each describing a control neural network consisting of a group of prototypical excitatory pyramidals cells and a group of prototypical inhibitory interneurons connected via of excitatory and inhibitory synapses. A nonlinear phenomenon in this neural network has been presented.

The brain is a complex controlled dynamical system with distributed parameters that can display behavior which is periodic or chaotic of varying dimensionality [7, 3, 40]. The controlled chaotic behavior in the brain is healthy whereas lower–dimensional chaos or periodicity is an indicator of disease. In the case of epilepsy, the analysis of bilinear model shows a seizure where many neurons are synchronized. If populations of neurons are communicating quickly among the nodes of the lattice synchronize isssss a periodic limit cycle. The bilinear lattice has complex controlled dynamics that should correspond to healthier neural tissue. These dynamics is desirable in brain activity because a chaotic state corresponds to an infinite number of unstable periodic orbits which would then be quickly available for neural computation [11, 27].

Acknowledgments

Research was partially supported in part by the NATO grant (SST.CLG 975032), by the NSF grants DBI-980821, EIA-9872509, and NIH grant R01-NS-31451

References

[1] Abraham, R., and Marsden J. (1978), *Foundation of Mechanics*, Benjamin–Cummings. Reading, MA.

[2] Arnold, V. (1974), *Mathematical methods of classical mechanics*, Springer–Verlag, New–York.

[3] Belair J., Glass L., an der Heiden U., and Milton J. eds. (1995), *Dynamical disease - Mathematical Analysis of Human Illness*, American Institute of Physics Press, Williston.

[4] Brockett, R. (1973), Lie algebras and Lie groups in control theory, in *Geometric mehods in system theory (Proc. of the NATO Advanced Study Institute held at London, August 27 September 7*, Mayne, D. and Brockett, R. (eds.), D. Reidel Publishing Company, Dordrecht–Holland/Boston.

[5] Butkovskiy, A. and Samoilenko, Yu. (1990), *Control of quantum-mechanical processes and systems*, Kluwer Academic Publishers, Dordrecht.

[6] Crouch, P. (1985), Variation characterization of Hamiltonian systems in *IMA J. Math. Cont. and Inf.* Vol. 3, 123–130.

[7] Destexhe A., Mainen Z. and Sejnowski T.J. (1994), Synthesis of models for excitable membranes, synaptic transmission and neuromodulation using a common kinetic formalism. *J. Computational Neurosci.* 1: 195-230.

[8] Dirac, P. (1958), *The principles of quantum mechanics*, Clarendon Press, Oxford.

[9] Dubrovin, B., Novikov, S., and Fomenko, A.(1984), *Modern geometry–Methods and Applications, Part. 1*, Graduate text in mathematics, Vol. 93, Springer–Verlag, New York.

[10] Feynman, R. (1982), Simulating physics with computers, *Int. Journal of Theoretical Physics*, Vol. 21, 467–475.

[11] Feynman, R. (1985), Quantum Mechanical Computers, *Optics News*, Vol. 11, 11–20.

[12] Freeman, W. and Skarda, C. (1985), Spatial EEG patterns, non-linear dynamics and perception: The neo-Sherringtonian view, *Brain Research Reviews*, Vol. 10, 147–175.

[13] Freeman, W. (2000), *Neurodynamics: An Exploration of Mesoscopic Brain Dinamics*, Springer–Verlag, London.

[14] Gomatam, R. (1999), Quantum Theory and the Observation Problem , in *Reclaiming Cognition. The primaci of action, intention and emotion*, Núñez, R. and Freeman, W. (eds.), Imprint Academic, UK, 173–190.

[15] Haken, H. (1996), *Principles of brain funcioning: a sinergetic approach to brain activity, activity, behaviour, and cognition*, Springer–Verlag, Berlin, Heidelberg, New York.

[16] Hendricson, A. (1995), The molecular problem: Exploiting structure in global optimization *SIAM J. Optimization*, Vol. 5, 835–857.

[17] Hiebeler, D. and Tater, R. (1997), Cellular automata and discrete physics *Introduction to Nonlinear Physics*, Lui Lam (ed.), Springer, New York–Berlin, 143–166.

[18] Hodgkin, A. and Hukley, A. (1952), The components of membrane conductance in the giant axon of Lofigo, in *J. Physiol.* , Vol. 116, 473–496.

[19] Hopfield, J. and Tank, D. (1985), "Neural" computation of decision in optimization problems *Biol. Cybern.* Vol. 52, 141–152.

[20] Hopfield, J. (1994), Neurons, dynamics and computation *Phys. Today* Vol. 47, No.2, 40–46.

[21] Horst, R. and Pardalos, P. (1995), *Hadbook of Global Optimization*, Kluwer Academic Publishers, Dordrecht,

[22] Jibu, M. and Yassue, K. (1995), *Quantum Brain Dynamics and Consciouness: An introduction*, John Benjamin Publishing Company, Amstedam/Philadelphia.

[23] Jurdjevic, V. (1997), *Geometric Control Theory*, Cambridge University Press, Cambridge.

[24] Kandel, E., Schwartz, J., and Jessell, T. (1991), *Prinziples of Neural Science*, Elsevier, New York.

[25] Kaneko, K. (1993), The coupled map lattice, in *Theory and Application of Coupled Map Lattices*, Kanepo, K. (ed), John Wiley and Sons, Chicheste–New York–Brishbane–Toronto–Singapure, 1–50.

[26] Kaneko, K. (2001), *Complex system: chaos and beyond: a constructive approach with applications in life sciences*, Springer-Verlag, Berlin

[27] Kelso, J. A. S. (1995). *Dynamic Patterns: The Self-Organization of Brain and Behavior*. MIT Press, Cambridge, MA.

[28] Klimontovich, Yu. (1995), *Statistical theory of open systems*, Kluwer Academic Publishers, Dordrecht, Boston.

[29] Landau, L. and Lifshitz, E. (1976), *Mechanics*, Pergamon Press, Oxford, New York.

[30] Larner, R., Speelman,B. and Worth, R. (1997), A coupled ordinary differential equation lattice model for the simulation of epileptic seizures, *Chaos*, Vol. 9, 795–804.

[31] Lloyd, S. (1993), A potentially realizable quantum computer, *Science*, Vol. 261, 1569–1571.

[32] Loskutov, A. and Mikhailov, A. (1990), *Introduction to synergetics*, Nauka, Moscow (in Russian).

[33] Mahler, G. and Weberruss, V. (1995), *Quantum networks : dynamics of open nanostructures*, Berlin, Springer.

[34] Marcus, L. (1973), General theory of global dynamics, in *Geometric Methods in System Theory*, Mayne D.Q. and Brockett, R.W. (eds.), D. Reidel Publisching Company, Dodrecht–Boston, 150–158.

[35] Melnikov, V. (1963), On the stability of the center for time-periodic perturbations *Trans. Moscow. Math. Soc.*, Vol. 12, No. 1, 1–56.

[36] Morris, C. and Lecar, H. (1981) Voltage oscillations in the barnacle giant muskle fiber, *Biophys. J.*, Vol. 35, 193–213.

[37] Pardalos, P., Liu, X., and Xue, G. (1997), Protein conformation of a lattice model using tabu search, *Journal of Global Optimization*, Vol. 11, No. 1, 55–68.

[38] Pardalos, P., Floudas, C., and Klepeis, J. (1999), Global Optimization Approaches in Protein Folding and Peptide Docking, in *Math. Sup. for Molec. Biol.*, DIMACS Series, Vol. 47, Amer.Math. Soc., 141–171.

[39] Ray, D. (1992), Inhibition of chaos in bistable Hamiltonian systems by a critical external resonances, *Phys. Rew. A*, Vol. 46, No. 10, 5975–5977.

[40] Sackellares, C., Iasemidis, L., Shiau, D., Gilmore, R. and Roper, S. (2000), Epilepsy — when chaos fails, in *Chaos in Brain?*, Lehnertz, K., Arnold, J, Grassberger, P. and Elger, C (Eds.), World Scientific, 112–133.

[41] Samoilenko, Yu. and Yatsenko, V. (1991), Quantum Mechanical Approach to Optimization Problems, in *International Conference on Optimimization Techniques, Vladivostoc*, Institute of Control Problems, Moscow.

[42] Traub, R., Miles, R. and Jeffreys, J. (1993), Synaptic and intrinsic conductances shape picrotoxin–induced synchronized after discharges in the quinca pig hippocampal slice, *J. Physiol. (London)*, Vol. 461, 525–547.

[43] Van der Shaft, A. (1982), Controllability and observability for affine nonlinear Hamiltonian systems, *IEEE Trans. Aut. Cont.*, Vol. AC–27, 490–494.

[44] Xue, G., Zall, A. and Pardalos, P.(1996), Rapid evaluation of potencial energy functions in molecular and protein conformation, in *DIMACS Series*, Vol. 23, Amer. Math. Soc., 237–249.

[45] Yatsenko, V. (1993), Quantum Mechanical Analogy of Bellman Optimal Principle for Control Dynamical Processes, *Cybernetics and Computing Techniques*, Vol. 99, 43–49.

[46] Yatsenko, V. (1995), Hamiltomian model of a transputer type quantum dynamaton, in *Quantum Communications and Measurement*, Plenum Publishing Corporation, New York.

[47] Yatsenko, V. (1996), Determining the characteristics of water pollutants by neural sensors and pattern recognition methods, *Journal of Chromatography*, Vol. 722, No. 1+2, 233–243.

Chapter 13

COMPUTATIONAL METHODS FOR EPILEPSY DIAGNOSIS. VISUAL PERCEPTION AND EEG.

Rasa Ruseckaite

Centre for Visual Sciences, RSBS, ANU

Canberra, Australia

Tel: +61 2 6125 4242

Fax: +61 2 6125 3808

Ruseckaite@rsbs.anu.edu.au

Abstract   This paper describes a preliminary algorithm performing automated epilepsy diagnosing by means of computational visual perception tests and digital electroencephalograph data analysis. Special machine learning algorithm and signal processing methods are used. The algorithm is tested on real data of epileptic and healthy persons that are treated in Neurology Clinics, Kaunas Medical University, Lithuania. The detailed examination of results shows that visual perception testing and automated EEG data analysis could be used for epilepsy diagnosing.

Keywords: Machine learning, visual perception, epilepsy, EEG, signal processing

1.   Introduction

According to the data of the 4th International Congress of Epileptology, Florence, 2000, 1 percent of all the persons in the world suffer from epilepsy. The 5 percent of them have the refractory epilepsy, when medicine is not able to treat such persons. Epilepsy is characterized by intermittent paroxysmal rhythmic electrical discharges that could disrupt normal brain functions. The most popular is the temporal lobe epilepsy.

People, suffering from it are totally discomforted. It is very important to diagnose this disease in time. Physicians use manual testing and diagnosing tools, but the results are not always correct. Sometimes it is impossible to diagnose and treat it precisely. Many computer specialists try to help doctors

P.M. Pardalos and J. Principe (eds.), Biocomputing, 225-241.
© 2002 *Kluwer Academic Publishers. Printed in the Netherlands.*

using various techniques for data processing. New methods are going to be investigated for the epilepsy prediction.

My goal was to create special methodic that could help to solve the problem, mentioned above. Sometimes happens, that people, having epilepsy, tend to have visual perception disorders as well. Visual perception disorders (known as visual agnosia) is persons' disability to distinguish among different colors and shadows, sizes, figures, shapes and many other objects. The visual perception tests were used for patients' diagnosis information. People, having epilepsy, also had some visual perception abnormalities (color perception disorders, size-figures discrimination problems, and others).

We have created a computerized system, which could be able to test epileptic persons' visual perception and predict epilepsy according to the visual perception testing results [2].

The created system also integrates the digital EEG (electroencephalography) analysis, used for preliminary epilepsy diagnosis

Epileptic persons were tested by means of Size - Form Discrimination and Munsell color test. Special machine learning (ML) techniques were used for testing data processing and creation of regularities. In parallel, special auto-regressive (ARMA_MULTI) methods were used as well.

Digital EEG processing was also used for the same task.

2.   Visual perception tests

2.1.   Farnsworth - Munsel (FM) 100 - hue test

The Farnsworth-Munsell 100-hue test was described by Farnsworth (1943) in terms of a color circle arranged on a uniform chromaticity diagram. There are 85 "chips" or color samples, divided into four quadrant-sets. The 85 were chosen from a series of 100 (hence the name) with Value 5 and Chroma 4, based on the data obtained by Nickerson and Granville in 1940. The 15 chips that were discarded came from 15 pairs of chips often confused by normal subjects. Therefore the 85 chips represent just-noticeably-different steps for subjects with normal color vision.

The Farnsworth- Munsel 100 - hue (FM) test [6] has been used for such purposes:

For examination the relationships between deficits in color and contrast discrimination in epileptic people;

To show that impaired colour discrimination is related to patho-physiology of various brain diseases and can be a sensitive method for early detection and monitoring of epilepsy.

For testing of colors discrimination, measuring the zones of colors confusion and learning programs of rehabilitation, the computerized Farnsworth - Munsell

100 - hue test was created and applied to test patients with various types of brain damage which was verified by means of EEG, CT or MRI.

The investigated person has to sort the colors according their shadows. Four basic colors are presented: red, green, blue and yellow.

Both of testing results are stored to the special database for their further processing.

2.2. Size - Form Discrimination (SFD) Test

The test procedure consists of sequential presentation of 100 pairs of vertically centered geometrical figures of variable size. The investigated person is given short amount of time to perceive and to compare each pair. Afterwards, he is asked to tell whether the left and the right figures were equal or not.

All figure pairs are clustered into 20 groups containing 5 pairs each. The left figure size is the same for all pictures of one group [1,5]. The right figure size is adjusted to be significantly less, less, equal, greater and significantly greater than the left figure. It is calculated as presented in the (1) expression. The (2) expression is used for the relative size of the left figure, which changes from 20 to 248 display units for different groups.

$$T = 20 + 12 * i, i = 0...19. \tag{1}$$

$$K = T + j * 5, j = -2..2 \tag{2}$$

Before launching the test supervisor chooses preferred test options. The shapes of both figures are not necessarily the same and are selected form the following set: square, circle, triangle, vertical line, and horizontal line. Each figure and the background are assigned one of 16 possible colors (black and white included). In addition, the figures can be chosen to be solid or hollow.

3. Data interpretation methods

3.1. The Nearest Neighbour Classifier

The Nearest Neighbor (*NN*) Classifier [10] is the classical method, which is used for data classification. The main principle of it is to find the minimum distance between classified data and the given class centers. All the distances are calculated, and the minimum of them describes the possible class, to which the person belongs.

At first the *NN* should be learned with the given apriori data in order to find the class centers. The class center can be expressed in the following way:

$$\overline{X_l} = \begin{pmatrix} x_{1,l} \\ x_{2,l} \\ ... \\ x_{N,L} \end{pmatrix} \quad ,(i = 1,...N; l = 1...L), \tag{3}$$

where $N = 100$ is the number of features, (SFD tests answers);
$l = 1..L, L = 6$ is the number of classes.

The $\overline{x_{i,L}}$ is one of the class center coordinates.

$$\overline{x_{i,L}} = \frac{1}{N} \sum_{k=1}^{N} x_{ikL} \quad ,(i = 1,...L; k = 1...K), \tag{4}$$

where K is the ammount of patients' data. x_{ikL} is one of the SFD (or FM) test set answer.

Consider, that $x \in \Omega_r$, where x is the patient' data, to be classified, belongs to class Ω_r, where Ω_r is the class (r=1..R = 6). We calculate the distances $d_1..d_6$ between sample and class centers. The minimum values among $d_{\min} = \arg\min_{l=1,2,...L} d_l$ of them shows that x belongs to class r.

3.1.1 Experiment.   The algorithm effectiveness was tested twice:

With the FM test results. Six different classes were used for training (ET, EO, EP, EF, HEALTHY and MTR (traumatic persons class). Unfortunately, the classification results were bad. The 25 patients' data were used for NN training and 25 – for recognition. Only 10 persons' diagnoses were recognized correctly. According to this, we decided to refuse using of FM testing results for diagnosis prediction.

The classification results were significantly better. NN classified about 60 percent of data correctly. The classification results are presented in Figure 13.1.

3.2.   CHARADE

Our system incorporates a simplified and slightly modified version of symbolic machine learning algorithm *CHARADE* [4]. This algorithm generalizes over pre-classified training examples inducing set of classification rules. In our case, the rule set corresponds to detected regularities and relates SFD test results to medical diagnosis.

3.2.1   Parameter space transformation.   In order to improve the interpretability of constructed rules, the mapping of all original 100-dimensional person's response vectors into the new parameter space precedes the induction process. Given the test data, the value of each new parameter corresponds to

Patients

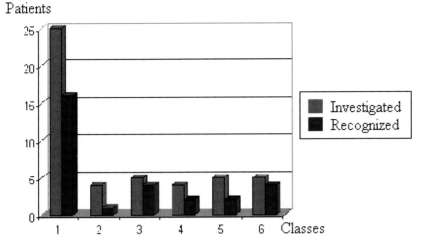

Figure 13.1.   NN efectiveness for SFD testing results

Table 13.1.   Partition of SFD test answers into subsets

	Small	Medium	Large	(Left Fig.)
Sgnf. Less	$A_{1,1}$	$A_{2,1}$	$A_{3,1}$	(size)
Less	$A_{1,2}$	$A_{2,2}$	$A_{3,2}$	
Equal	$A_{1,3}$	$A_{2,3}$	$A_{3,3}$	
Greater	$A_{1,4}$	$A_{2,4}$	$A_{3,4}$	
Sgnf. Greater	$A_{1,5}$	$A_{2,5}$	$A_{3,5}$	
(right fig.size)				

the percentage of correct answers within a specified set of test answers. We have defined and used 72 overlapping answer sets $A_{i,j}$ ($i=1...6$, $j=1,...,12$) (Table 13.1).

$$A_{i,6} = A_{i,1} \cup A_{i,2} \quad A_{i,7} = A_{i,4} \cup A_{i,5}$$
$$A_{i,8} = A_{i,1} \cup A_{i,5} \quad A_{i,9} = A_{i,2} \cup A_{i,4}$$
$$A_{i,10} = A_{i,2} \cup A_{i,3} \cup A_{i,4}$$
$$A_{i,11} = A_{i,1} \cup A_{i,2} \cup A_{i,4} \cup A_{i,5}$$
$$A_{i,12} = A_{i,1} \cup A_{i,2} \cup A_{i,3} \cup A_{i,4} \cup A_{i,5}$$
$$A_{4,j} = A_{1,j} \cup A_{2,j} \quad A_{5,j} = A_{2,j} \cup A_{3,j}$$
$$A_{6,j} = A_{1,j} \cup A_{2,j} \cup A_{3,j} ,$$

3.2.2   Induction results.

We obtained two different rule sets. The first set covered all epilepsy forms, discriminating them from the rest of diagnoses (including healthy). The second rule set discriminated between possible epilepsy variations (temporal, frontal, occipital, parietal). Each collection contained tens or even hundreds of rules depending on learning algorithm settings. Thus, we present only a few examples taken from both rule sets:

– if ($C_{5,5}$ $\leq 39\%$) then diagnosis = epilepsy.

– if ($24\% \leq C_{3,12}$ $\leq 43\%$) then diagnosis = epilepsy.

– if ($C_{4,11} > 68\%$) $\cup$ ($C_{3,1} > 67\%$) $\cup$ ($33\% \leq C_{3,3}$ $\leq 67\%$) then diagnosis = other.

– if ($C_{6,5} > 68\%$) $\cup$ ($C_{2,9} > 67\%$) $\cup$ ($42\% \leq C_{4,12}$ $\leq 62\%$) then diagnosis = other.

– if ($C_{1,4} > 71\%$) $\cup$ ($33\% \leq C_{5,3}$ $\leq 67\%$) then diagnosis = temporal epilepsy.

– if ($C_{6,3}$ $\leq 32\%$) $\cup$ ($63\% \leq C_{2,12}$ $\leq 86\%$) $\cup$ ($38\% \leq C_{3,7}$ $\leq 69\%$) then diagnosis = frontal epilepsy.

– if ($C_{6,3}$ $\leq 32\%$) $\cup$ ($C_{2,1}$ $\leq 33\%$) $\cup$ ($42\% \leq C_{3,12}$ $\leq 60\%$) then diagnosis = occipital epilepsy.

– if ($C_{3,8}$ $\leq 35\%$) $\cup$ ($33\% \leq C_{5,3}$ $\leq 67\%$ ($\cup$ ($30\% \leq C_{1,10}$ $\leq 43\%$) then diagnosis = parietal epilepsy.

Here, $C_{i,j}$ means correct answer percentage in $A_{i,j}$ answer set.

3.2.3   The effectiveness of CHARADE.   In order to try the *CHARADE* effectiveness:

The algorithm was learned by SFD testing results of five basic classes.

1. ET (patients, having temporal lobe epilepsy)

2. EO (patients, having occipital lobe epilepsy)

3. EF (patients, having frontal lobe epilepsy)

4. EP (patients, having parietal lobe epilepsy)

5. HEALTHY (healthy persons)

After that the algorithm was tested with other data of the same classes. We tried to find out how many diagnoses *CHARADE* recognizes correctly. The results are presented in Figure 13.2.

The algorithm was learned using 13 classes, when the basic five ones were diveded into subclasses. For example, ET was divided into ETS (temporal lobe epilepsy of right brain part), ETD (temporal lobe epilepsy of left brain part), EU (temporal lobe epilepsy of the both brain parts). The learning and recognition procedure was the same as for five classes. The results are presented in Figure 13.3.

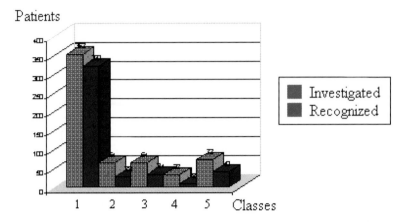

Figure 13.2.   CHARADE efectiveness for basic five classes

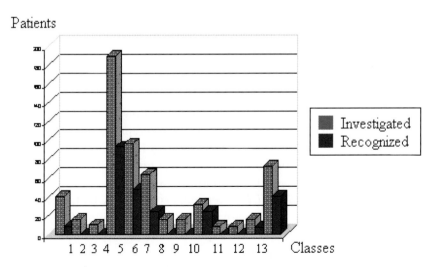

Figure 13.3.   CHARADE efectiveness for 13 classes

According to the received data, we can state that *CHARADE* is more effective, when it is learned within five basic classes data. Without dividing them to subclasses, it is possible to receive about 70 percent effectiveness.

3.3.   Regressive methods

The *CHARADE* and *NN* system results were positive and showed 70 percent corectiviness. Anyway, it was not sufficient for good diagnose detection and creation of regularities. Some of the rules overloaded, the finite interpretation was not informative in some cases (it was stated that person can have temporal as well as parietal epilepsy, while those diagnoses are totally separate).

In comparison with the mentioned above methods, the AR method [7, 8] had been chosen. This method is able to process many factors (features) in a time as well. This algorithm was very good in currency rate prognosing.

The *ARMA_MULTI* model is the AR model, capable with many factors in a time. The *ARMA_MULTI* model could be present in the following expression:

$$w_t = \sum_{i=1}^{p} a_i w_{t-i} + \sum_{j=1}^{q} b_j \varepsilon_{t-j} + \varepsilon_t \tag{5}$$

The first expression is auto-regressive part of model, and the second - is the multi- average sum. Value w_t describes the real mean of patient's diagnose. ε is the error rate. It can be calculated between real and predicted diagnose value. The a and b expressions are *ARMA_MULTI* parameters.

3.4.   ARMA_MULTI algorithm

ARMA_MULTI algorithm can be presented as follows:

The test data set at the $t = to$ time moment is being preprocessed. *ARMA MULTI* is acknowledged, using already existing test data (Munsell color perception testing results and size – form discrimination test results). *ARMA MULTI* parameters a and b are calculated in the following way:

I used simplified *ARMA_MULTI* version, without MA part, not working with data multi averaging. In my case parameter a is the parameter, presenting the feature dependency on patient's diagnose. The parameter b is set as zero (because of escaping of MA part).

$q = 0$ because of $b = 0$. $p = 1$ (the prognose does not depend on the patient's number in the data sequence). All parameters are minimized:

$$\min_{b}\min_{a} \sum \varepsilon_t^2(a, b) \quad => \quad \min_{a} \sum \varepsilon_t^2(a) \tag{6}$$

Table 13.2. ARMA_MULTI working results

D	I	ET	EF	EO	EP
EO	ET	1.38e-01	-6.72e-02	-2.18e-02	-2.24e-01
EP	EP	2.39e-01	9.37e-02	2.25e-01	2.76e-01
ET	ET	1.393e-01	-2.38e-03	-2.17e-01	-4.07-e-02
ET	EP	-3.66e-01	-1.58e-01	-9.75e-02	-1.234e-01
ET	ET	1.673e-01	-1.34e-03	-1.17e-01	0.07-e-02
ET	ET	1.6e-01	-1.38e-03	-2.8e-01	-2.07-e-02
ET	ET	2. 93e-01	-1.08e-03	-1.4e-01	-2.04-e-02
ET	ET	4.43e-01	2.678e-03	-2.188e-01	1.07-e-02
ET	ET	3.383e-01	4.38e-03	-1.67e-01	0.17-e-02
EF	EF	-2.74636e-02	2.7038e-01	1.10745e-01	1.59746e-01
HL	ET	7.57677e-02	-4.3702e-02	-8.9990e-02	9.32991e-02

$$a(b) = \arg\min_{a} \sum_{t=t_0}^{T} \varepsilon_T^2(a,b), \text{where} b = 0 \qquad (7)$$

$$\min_{b} \sum \varepsilon_t^2(a,b), b), \text{where} b = 0 \qquad (8)$$

Risk function is calculated. *ARMA_MULTI* works with $m = 19$ separate factors [8,10] and prognoses four separate classes: $\{a_i^{(j)}, \varepsilon_i^{(j)}\}, j = 1, ..N$, where $N = 6$ is the number of classes

The first four columns are used for patient's brain state: $p_1^{(i)} = \{0,1\}$, where l - basic factor number, $l = 1..6$, o $i=1..6$, and i – state number.

Factor can be:

0 when a patient is healthy;

1 when a patient has epilepsy.

The rest columns serve for features $p_1^{(i)} = \{-1,0,1\}$, where l - number, $l = 5..100$, o $i = 1..6$, and i – state number. Each of them can be: -1,0 or 1 valued. ARMA_MULTI worked with linear visual perception test data.

They were not transformed in the parameter space. The four classes (brain states) were prognosed: Temporal epilepsy (ET); frontal epilepsy (EF); ocipital epilepsy (EO); parietal epilepsy (EP).

ARMA_MULTI was capable to present so called "likelihood" values, which described the potential diagnose (Table 13.2).

The second column is for real diagnose, the third – for *ARMA_MULTI* interpretation. The rest columns shows the "likelihod" diagnose value. Figure 13.4 shows the *ARMA_MULTI* working results.

Here you can see that the class 1 (ET) – which has the biggest ammount of samples, is not recognized very well. Only 1/3rd of diagnoses are interpretted correctly.

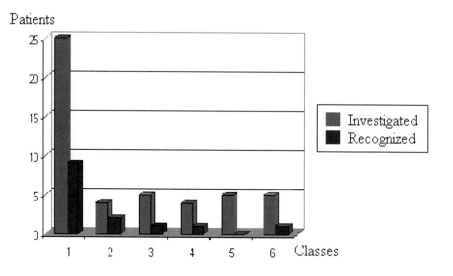

Figure 13.4. ARMA_MULTI efectiveness

According to the obtained data, it could be stated that *CHARADE* was the most effective method for VPD classification.

I have already mentioned that *ARMA_MULTI* is not so good as *CHARADE*, which created rules and regularities can better predict patients diagnose. I can state that AI methods are more productive in the concrete situation.

Three possible reasons, which can tend a bad *ARMA_MULTI* prognosis, are listed below:

Not sufficient amount of patients' data for *ARMA_MULTI* acknowledgement or too many factors, used by *ARMA_MULTI*.

Diagnose values: patient is ill, or not. The values have a Boolean expression: $\{1, 0\}$.

CHARADE is being modified at the moment and is going to be used for secondary prognoses and for other data as well. I am planning to include digital EEG data for epilepsy prognosis as well.

The *NN* classifier was also ineffective and could not be used for epilepsy prediction.

It was decided to combine the *CHARADE* rules and other medical data.

4.   EEG analysis

The EEG is the most effective and useful method for epilepsy diagnosing. It was decided to combine *CHARADE* rules and digital EEG analysis for epilepsy diagnosis.

I worked with three groups of persons - epileptic persons (EPI), traumatic persons (MTR) and healthy people (HEALTHY). I tried to ascertain whether some EEG data regularity exists for those three groups.

Three EEG realizations of 16-channel digital EEG were used. Realizations were as follows:

Background (closed eyes),

Hyperventilation

Photo stimulation EEG.

The whole EEG process can be expressed as matrix, that contains $m = 16$ rows (channels) and N columns (signal lenghts in seconds).

$$Y = \begin{bmatrix} y_1^1 & y_1^2 & \cdots & y_1^N \\ y_2^1 & y_2^2 & \cdots & y_2^N \\ \vdots & \vdots & \cdots & \vdots \\ y_m^1 & y_m^2 & \cdots & y_m^N \end{bmatrix}, m = 16, \tag{9}$$

where m - the number of channels;

$t-$ the realization lengths (time).

Every of 16th processes could be presented as follows:

$$y_t = \begin{bmatrix} y_l^t, y_l^t, \dots y_L^T \end{bmatrix}, \quad t = 1..N; \quad l = 1..L = m; \tag{10}$$

The experiment was performed using special MATLAB software, which is very convenient for signal processing and has a lot of possibilities for data processing. Two ways of EEG processing were chosen: spectral and LPC (linear prediction coding) [3] analysis.

4.1. Spectral EEG analysis

The algorithm is presented below:

Three EEG realizations were analyzed for every of EEG 16 channels (see above).

Those 16 channels were split to 15 equal parts along the realization length. The lengths of every part is t = 3 sec.

Fast Fourie transformation was calculated for EEG data part, received in step 2.

$$f(s) = \int f(x)exp(-i2xs)dx \tag{11}$$

After that 15*16 dimensions signal was received. (For every EEG data)

16 spectral characteristics (maximums for frequencies and amplitudes) for three classes (EPI, MTR and HEALTHY) were calculated.

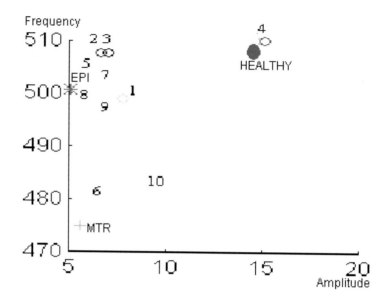

Figure 13.5.   Classification results for three classes

$$f_i(s) = \frac{1}{N} \sum_{i=1}^{N=15} f_i(s) \tag{12}$$

Spectral maximums were calculated for every given EEG realization

$$s_{\max} = \arg\max \ f_i(s), \ 0 \le s \le s_{\max} \tag{13}$$

30 persons' data were used for algorithm training and 10 patients' data – for algorithm testing. My task was to clarify, whether this algorithm can classify correctly persons' data according their spectral EEG characteristics. The Euclid's [6] classifier was chosen.

The classification results are presented below, in Figure 13.5.

Table 13.3 shows the results for 10 persons according their spectral maximums for EEG background realization. These persons had the temporal lobe epilepsy. Seven of them were classified correctly, two of them - as MTR, and one – as HEALTHY.

4.2.   EEG LPC parameters analysis

In order to solve the problem, the LPC method was chosen. It is also known as Durbin's method [3] and can formally be given as the following algorithm.

Table 13.3.   EEG Interpretation after spectral analysis

Pat./Realization	Background	Photostymulation	Hippervent.
Patient 1	EPI	MTR	-
Patient 2	EPI	EPI	EPI/MTR
Patient 3	EPI	MTR	MTR/EPI
Patient 4	HLT	-	-
Patient 5	EPI	EPI	-
Patient 6	MTR	EPI/MTR	EPI
Patient 7	EPI	EPI	EPI
Patient 8	EPI	EPI	-
Patient 9	EPI	MTR	MTR/EPI
Patient 10	MTR	MTR	EPI

1. The EEG process is a random one and can be described as follows:

$$X_t = -\sum_{i=1}^{p^{(1)}} a_i^{(1)} X_{t-i} + b^{(1)} v_t \tag{14}$$

where $\theta = (p, b, a_1, ..., a_p)$, and $v_t \sim N(0, 1)$ are independent random values.

2. While the EEG data are different for different brain lobes, it is possible to find the separate LPC parameters for them.

3. My task was to find the parameters a and b for classes $r = 3$, (EPI, MTR and HLT) which could describe the class of EEGdata x_t and could classify it to one of the given classes, as follows:

$$\begin{aligned} x_t \in \Omega_r, \; (r = 1, \, 2, \, ..., \, M), \\ x_t \; (t = \; 1, ..., n) \end{aligned} \tag{15}$$

4. The parameters a are calculated as the (16) and (17) expressions present:

$$\alpha_j^{(i)} = \alpha_j^{(i-1)} - k_i \alpha_{i-j}^{(i-1)} \tag{16}$$

$$a_m = LPC\, parameters = \alpha_m^{(p)}, \quad 1 \le m \le p, \quad p = 10 \tag{17}$$

The algorithm was trained within the data of three (EPI, MTR and HEALTHY) classes. LPC parameters were calculated for every process, and for every class. The Euclid's classifier was used for the algorithm efectiveness detection.

Patients

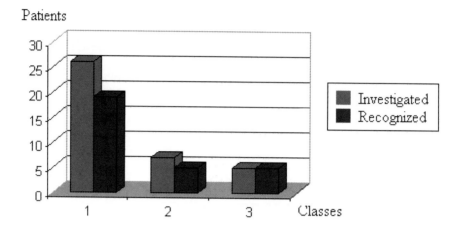

Figure 13.6.   LPC Classification results for three classes

It was trained with the 40 persons data, and was tested with the 40 persons data as well.

Figure 13.6 shows the classification results. We see that about 80 percent were classified correctly.

It is possible to state that LPC analysis method is more effective than spectral analysis one.

5.   LPC and CHARADE interpretation

According to the results, it was decided to combine both EEG LPC and CHARADE methods at the same time.

The results are presented in Figure 13.7.

The digital EEG analysis showed that, can be LPC methods could be used together with machine generated rule sets (*CHARADE*) for epilepsy prediction. Both combined methods showed the 89 percent efectiviness.

According to the efectiveness, it was decided to propose the model for the automated epilepsy diagnosis. The model is presented in Figure 13.8.

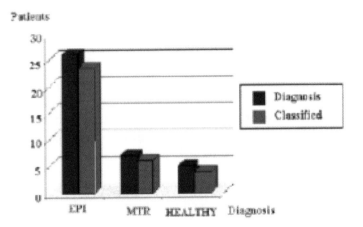

Figure 13.7.   Combined LPC and Charade Classification results for three classes

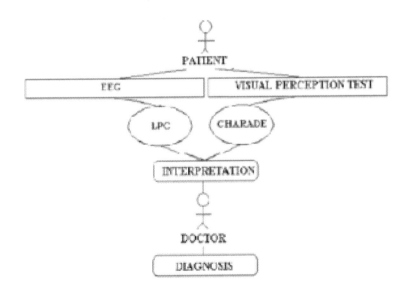

Figure 13.8.   Automated epilepsy diagnosing system model

6.   Conclusions

We have developed the computer system that encompasses together VP test procedure, symbolic machine learning algorithm and digital EEG processing methods.

Our machine learning subsystem was capable of extracting interesting dependencies that related VP test results to person's diagnosis. Even if induced rules show near perfect accuracy on the learning set examples additional data and experiences are required for being able to conclude that machine generated rule sets ate appropriate for computer aided diagnosing based on VP test.

We have realized, that it is useful to combine both visual perception testing data and EEG processing methods. CHARADE and LPC gave the best classification results. The sensitivity for the classified data was about 89 percent. Only 11 percent of given samples were misclassified.

The results show, that visual perception factor (size and discrimination) could be used for early epilepsy detection stage.

References

[1] Bulatov, A., and A. Bertulis (1994). A neuropsychological model of the Ophel Kundt illusion perception, Vol Supplement ECVP'94, Eindhoven, Abstracts, 56, p.46-47

[2] Christianson, A (1992). Visual half-field testing of memory functions in patients considered for surgical treatment of intractable complex partial epilepsy, Acta neural Scand, pp. 86.

[3] Jenkins, G., and D. Watts (1989). Spectral Analysis & Its Applications (1,2), Moscow, Mir (in Russian).

[4] Ganascia, J.G.(1987). Charade: a rule system learning system, IJCAI'87, p.234-239.

[5] Godefroid, J. (1988). Les chemins de la psychologie, 1, Pierre Mardager., pp.45-85.

[6] Lukauskienà R, V. Vilūnas, and K. Gurevičius (2000) Computerized program for investigation and diagnosis of colour perception, Proceedings of International Conference Biomedical Engineering, Kaunas, pp. 55-59

[7] Mockus J., and L.Mockus (1991). Bayesian Approach to Global Optimization and Applications to Multi-Objective Constrained Problems, Journal of Optimization Theory and Applications, , vol 70, No.1, p. 155-171.

[8] Mockus J., (2000). A set of examples of global and discrete optimization, Kluwer

[9] Ruseckaite, R.. G.Raškinis, R.Lukauskienà (1997). Computer interactive system for ascertainment of visual perception disorders, Proceedings of international conference Biomedical Engineering, Kaunas, p. 178-181

[10] Ruseckaite R. (2000). Epilepsy prediction by means of artificial intelligence methods, Proceedings of international conference Biomedical Engineering, Kaunas, p. 61-66

Chapter 14

HARDNESS AND THE POTENTIAL ENERGY FUNCTION IN INTERNAL ROTATIONS: A GENERALIZED SYMMETRY-ADAPTED INTERPOLATION PROCEDURE

Cristian Cardenas-Lailhacar

Industrial and Systems Engineering,
University of Florida
303 Weil Hall, P.O. Box 116595,
Gainesville, FL 32611-6595

Abstract   Density functional theory (DFT) has provided the theoretical basis to study re-activity parameters of molecular systems, this is, to its resistance to change its electron density distribution, namely the molecular hardness (η). Its evolution along a given reaction coordinate is examined in order to better understand the chemical reactivity of the system of interest. In this work, compounds that are often used as prototypes to study their linkage in proteins are shown.

In the DFT framework, for which the total number of electrons must be con-served, rotational isomerization reactions can be regarded as a reorganization / redistribution of electron density among atoms in a system.

The relation between the molecular hardness and the potential energy function (E) (both global properties) is here studied. It is shown that η is analytically related to E. Correction terms to the definition for η, that include a molecular symmetry parameter (b), are included. An expression for the activation hardness in terms of the activation potential energy function and b for the molecular system under study are derived. A qualitative proof of the principle of maximum hardness (PMH) is shown. The procedure is tested with nine molecular systems. Our results show to be in excellent agreement with those available in the literature.

1.   Introduction

Isomerization reactions occur in a large number of molecular systems that are present in nature, as well as other man-made ones. In particular, simple molecular systems that undergo internal rotation, like trans-cis isomerization reactions, are of main interest as they can provide insight to understand bigger and more complicated systems. Species containing sulfur atoms are usually

P.M. Pardalos and J. Principe (eds.), Biocomputing, 243-264.

used as prototypes to study linkage in proteins containing sulfur – sulfur bonds, and variations with hetero-atoms.

The purpose of this work is to derive general analytical expressions to better represent the torsional potential and hardness, and the relations among them. We will write the potential and hardness activations in terms of local reference (initial and final conformations) ionization potential and electron affinities.

The result is a general expression that allows their application to a broader number of systems, as it has built in the principal rotational axis (**b**) of the symmetry of the system point group of the reference conformations (initial and final). Although we have already derived expressions that give us great insight about the transition state from the potential energy stand point [1], we derive here additional expressions for the hardness, and its relation with the potential energy function. It is shown that previous works on the field are particular cases of the more general expressions that we will derive here. An expression for the activation hardness in terms of the activation potential energy function, and b, for the molecular system under study are derived. Hammond's postulate and its quantitative relation with the Brönsted coefficient ($\beta(b\alpha)$, is discussed in conjunction with particular cases. Finally, different symmetry group systems (i.e., different values of **b**) are studied to show the usefulness of the algorithm.

In this work we will concentrate on the hardness profile and the parameters that allows for its correlation to the potential energy function. In the next section we will introduce the necessary theoretical background for the potential energy and the hardness functions, and establish their relationship. Analytical expressions are obtained and the ratio of the forces involved is studied in conjunction with the potential and hardness at the TS. We continue with Hammond's postulate, the Bronsted coefficient, and their contributions to account for the isomerization processes described in this work. We conclude the theoretical treatment with the study of some special cases. Finally some applications are shown. The last section contains our conclusions.

2. Theoretical considerations

The Torsional Potential Function $E(b\alpha)$

In a previous work [1], we have shown that the potential energy function to describe an internal rotation can be written in terms of the torsional angle α and the principal proper rotational axis of the system point group symmetry b as follows:

$$E(b\alpha) = \Delta E^\circ \beta(b\alpha) + 2\omega_i(b\alpha)\omega_f(b\alpha)[K_i\omega_i(b\alpha) + K_f\omega_f(b\alpha)]/b^2 \qquad (1)$$

Where ΔE° is the difference in the potential energy between the reference conformations (initial and final states), and $\beta(b\alpha)$ is the Brönsted coefficient.

Here $\omega_i(b\alpha)$ and $\omega_f(b\alpha)$ are state weight functions [1] and K_i and K_f second derivatives of the potential function at the reference isomers. The Brönsted coefficient and the weight functions also change according to α and b:

$$\beta(b\alpha) = 1/2(b^2 + 1) - \omega_i(b\alpha)\{\omega_f(b\alpha)[2\omega_i(b\alpha) - 1] + b^2\}, \qquad (2\text{-}a)$$

where,

$$\omega_i(b\alpha) = [1 + \cos(b\alpha)]/2 \text{ and } \omega_i(b\alpha) + \omega_f(b\alpha) = 1 \qquad (2\text{-}b)$$

For convenience we will adopt the following notation: $\omega_i = \omega_i(b\alpha)$ and $\omega_f = \omega_f(b\alpha)$. In particular, for the potential function, and later on for the hardness function, its dependency on α is through the weight functions, consequently we will write: $E[\omega_{i,f}] = E[\omega_i(b\alpha)\omega_f(b\alpha)] = E(b\alpha)$. We shall return to the importance of the state weight functions later on when we discuss Hammond's postulate and the Brönsted coefficient.

It is clear from equation (1) that an estimate for the location of the Transition State (TS) can be derived from here by taking the first derivative of the potential with respect to the dihedral angle [1]. The second derivatives K_i and K_f, are obtained by computing two points in the catchment region of each reference state (initial and final). It has been shown [1] that with them the potential energy profile can be fully reproduced. In fact we suggest at the end of this work that this is true for any property that has a dependency on b and α. A particular case to any representation is also discussed later on.

Symmetric and Asymmetric Contributions

It is immediate, from equation (1) that the potential function can be written in terms of an asymmetric and symmetric contribution with respect to b and α [2], as follows:

$$E(b\alpha) = E_a(b\alpha) + E_s(b\alpha), \qquad (3\text{-}a)$$

where

$$E_a[\omega_{i,f}] = \Delta E°\beta(b\alpha) = \Delta E°\beta[\omega_{i,f}], \qquad (3\text{-}b)$$

and

$$\begin{aligned} E_s[\omega_{i,f}] &= 2\omega_i(b\alpha)\omega_f(b\alpha)[K_i\omega_i(b\alpha) + K_f\omega_f(b\alpha)]/b^2 \\ &= 2\omega_i\omega_f[K_i\omega_i + K_f\omega_f]/b^2. \end{aligned} \qquad (3\text{-}c)$$

This separation will help us to rationalize the hardness dependence on the PEF later on when we establish their relationship.

The DFT Hardness Function (η)

According to Hartree-Fock-Koopman's theorem, and for closed shell systems, the hardness (η) for a given system can be written in terms of its ionization potential (I) and electron affinity (A):

$$\eta \approx (I - A)/2 \text{ or } \eta = (\varepsilon_{LUMO} - \varepsilon_{HOMO})/2, \tag{4}$$

where

$$\varepsilon_{HOMO} = -I \text{ and } \varepsilon_{LUMO} = -A.$$

The hardness (η) is a scalar global property that can be written as a function that fulfils the symmetry elements (b) of the system and the reaction coordinate (α), provided that the total number of electrons remains constant.

Therefore, and as for the PEF [1, 2], the hardness can be written as a function of the dihedral angle and the symmetry parameter as follows:

$$\begin{aligned} \eta[\omega_{i,f}] &= \eta_a[\omega_i, \omega_f] + \eta_s[\omega_i, \omega_f] \\ &= \Delta\eta^\circ \beta'(b\alpha) + 2\omega_i\omega_f[L_i\omega_i + L_f\omega_f]/b^2 \end{aligned} \tag{5}$$

It has been shown [2] that both the potential and the hardness (HF) functions have the form of an asymmetric parabola with different curvatures. As a consequence, the location of the transition state for these two profiles is different. This is represented by different Brönsted coefficients ($\beta[\omega_{i,f}]$ for the PEF, and $\beta'[\omega_{i,f}]$ for the H). These considerations indicates that the potential energy and the hardness functions, as a function of the dihedral angle (α) and the symmetry parameter (b), undergo opposite behaviors. This is reproduced by our calculations as it will be shown later when we discuss our results. A key point of this work is to show that this behavior is kept for any value of b. In the special cases section. We will show that by considering $b = 1$, the results found in the literature are reproduced here as a particular case of this generalized symmetry-adapted interpolation procedure derived in this work.

We proceed by writing the ratio between the hardness and the potential energy functions (Eqns. (1) and (5)). This yields the following relation

$$\eta[\omega_{i,f}] = (\Delta\eta^\circ/\Delta E^\circ)E_a[\omega_{i,f}] + \frac{[L_i\omega_i + L_f\omega_f]}{[K_i\omega_i + K_f\omega_f]E_s[\omega_{i,f}]}$$

To be consistent with the literature we now define the parameter m_s as the ratio of the sum of the second derivatives of the HF and the PEF. This is:

$$m_s[\omega_{i,f}] = \frac{[L_i\omega_i + L_f\omega_f]}{[K_i\omega_i + K_f\omega_f]} = \frac{\eta[\omega_{i,f}] - \Delta\eta^\circ \beta'[\omega_{i,f}]}{E[\omega_{i,f}] - \Delta E^\circ \beta[\omega_{i,f}]} \tag{6}$$

Notice that m_s is a function of the reaction coordinate α and the symmetry parameter b through the state statistical weight functions. We will come back to this expression later on when we study the forces ratio. rearrange equation (6) to write an expression of the hardness in terms of the potential:

$$\eta[\omega_{i,f}] = m_s[\omega_{i,f}]E[\omega_{i,f}] - m_s[\omega_{i,f}]\beta[\omega_{i,f}]\Delta E^\circ + \beta'[\omega_{i,f}]\Delta\eta^\circ \quad (7)$$

In addition we introduce an asymmetric term (m_a) which is the ratio between the difference on the value of the reference states for the PEF and the HF:

$$m_a = \Delta\eta^\circ/\Delta E^\circ \quad (8)$$

We now use this relation to write a final expression of HF in terms of the PEF. This is:

$$\eta[\omega_{i,f}] = m_s[\omega_{i,f}]E[\omega_{i,f}] - \{(m_s[\omega_{i,f}]\beta[\omega_{i,f}]) - m_a\beta'[\omega_{i,f}]\}\Delta E^\circ \quad (9)$$

Notice that this expression does not suggest that have become independent of the symmetry term b as found in the literature. This term is included in the expression of the Brönsted coefficients $\beta[\omega_{i,f}]$ and $\beta'[\omega_{i,f}]$, and the potential energy function $E[\omega_{i,f}]$.

It should be pointed out that the expression that correlates the potential and the hardness functions, for the symmetric and asymmetric contributions, can also be written as:

$$\Gamma_a[\omega_{i,f}] = m_a[\omega_{i,f}]E_a[\omega_{i,f}] \quad (10\text{-}a)$$

$$\Gamma_s[\omega_{i,f}] = m_s[\omega_{i,f}]E_s[\omega_{i,f}] \quad (10\text{-}b)$$

These relations help to rationalize $\eta[\omega_{i,f}]$ along $E[\omega_{i,f}]$ as follows:

$$\begin{aligned}\eta[\omega_{i,f}] &= \Gamma_a[\omega_{i,f}] + \Gamma_s[\omega_{i,f}]\\ &= m_a[\omega_{i,f}]E_a[\omega_{i,f}] + m_s[\omega_{i,f}]E_s[\omega_{i,f}]\end{aligned} \quad (11)$$

This is a more general expression of the function shown in equation (9). The general form of the potential energy and the hardness functions is shown in Figure 14.1. Initial, final, and the Transition State are shown, as well as the difference among stable states and TS.

The Forces Ratio (m_s)

Although in many examples the second derivatives in the states cannot be considered as forces, in this work we will do so because we have shown that is their value that determines the shape, barrier and location of the TS. This parameter (m_s) is generally written and used as if the weight functions $[\omega_{i,f}]$

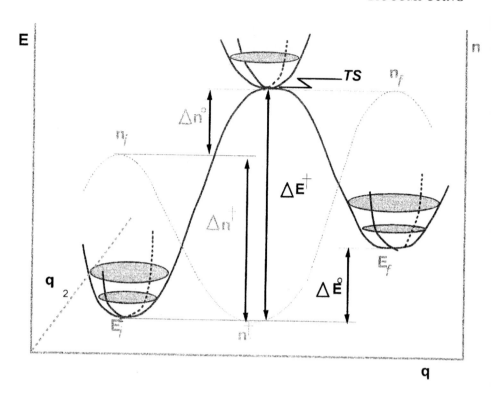

Figure 14.1. General form of the potential energy and the hardness functions. Initial, final, and transition state are shown, as well as the difference among stable states and TS.

were evaluated at the midpoint between the initial and final conformations, factoring out their values, this is, as a constant and only for cases where b = 1. In this work the expression found for m_s (equation (6)), shows that the results found in the literature becomes a particular case when b is set to unity. We shall see in our examples, that this value turns out to be the inflection point of $m_s[\omega_{i,f}]$. In Figure 14.2 we show the behavior of $m_s[\omega_{i,f}]$ as a function of the reaction coordinate for eight molecular systems for which $b = 1$.

To find the critical values of m_s we take its derivative against the reaction coordinate (α). This is:

$$\partial m_s[\omega_{i,f}]/\partial \alpha = 0,$$

which yields three possible values for m_s

$$m_{s-i} = \frac{L_i}{K_i} \qquad (12a)$$

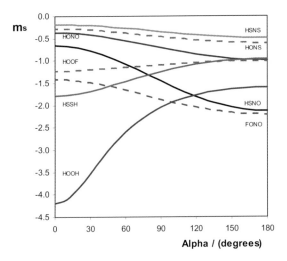

Figure 14.2. Evolution of $m_s[\omega_{i,f}]$ as a function of the reaction coordinate for eight molecular systems, for which $b = 1$.

$$m_{s-f} = \frac{L_f}{K_f} \tag{12b}$$

$$m_{s-TS} = \frac{[L_i + L_f]}{[K_i + K_f]} \tag{12c}$$

These results indicate the initial (m_{s-i}), final (m_{s-f}) and $TS (m_{s-TS})$ values of m_s at critical points, this is, minima, maxima and TS. Notice that, so far, these values do not determine if the PEF is typically bell shaped, or inverted bell shaped. An important result of this work is that the value of m_s at TS (m_{s-TS}), shown in the literature, is reproduced here (Equation (12-c)), showing again that those are particular cases of the general algorithm presented in this work. This can also be seen by considering Equation (6) with b = 1 and TS at $\alpha = \pi/2$ for the initial and final statistical weight functions; the result is Equation (12c). For those cases with different values of b and α the results will be numerically different, but similar in their evolution along the reaction coordinate because the periodicity and the symmetry of the system under consideration will vary such that these relations will hold.

Because of these results, we can certainly write expressions for the hardness as a function of the potential energy function that will better represent local (initial, final and TS) situations for the HF, by simply replacing the optimal values found in Equations (12)

$$\eta_{s-i}[\omega_{i,f}] = m_{s-i}\{E[\omega_{i,f}] - \beta[\omega_{i,f}]\Delta E^\circ\} + \beta'[\omega_{i,f}]\Delta\eta^\circ \tag{13a}$$

$$\eta_{s-f}[\omega_{i,f}] = m_{s-f}\{E[\omega_{i,f}] - \beta[\omega_{i,f}]\Delta E^\circ\} + \beta'[\omega_{i,f}]\Delta\eta^\circ \qquad (13b)$$

$$\eta_{s-TS}[\omega_{i,f}] = m_{s-TS}\{E[\omega_{i,f}] - \beta[\omega_{i,f}]\Delta E^\circ\} + \beta'[\omega_{i,f}]\Delta\eta^\circ \qquad (13c)$$

Equations (13a) and (13b) indicate that for a local representation the local forces are needed. This is, two points, two different conformations, need to be computed in the local regions. It is found in the literature that Equation 13c is widely used. A result reproduced in this work as a particular case of the general symmetry-adapted procedure. We will also show that this value is midway between the initial and final states (conformations), which will become clear when we discuss some examples.

Potential ($\Delta E^\dagger$) and Hardnes ($\Delta\eta^\dagger$) at the TS

The Potential Function at the Transition State (α_o) can be written in the general form shown in Equation (1):

$$\Delta E^\dagger(b\alpha_o) = \Delta E^\circ\beta(b\alpha_o) + 2\omega_i(b\alpha_o)\omega_f(b\alpha_o)[K_i\omega_i(b\alpha_o) + K_f\omega_f(b\alpha_o)]/b^2 \qquad (14)$$

As before, the symmetric and asymmetric contributions are written as:

$$E_a^\dagger[\omega_{i,f}^\dagger] = \Delta E^\circ\beta(b\alpha_o)$$

and

$$E_s^\dagger[\omega_{i,f}^\dagger] = \Delta E^\dagger - \Delta E \circ \beta(b\alpha_o) = 2\omega_i(b\alpha_o)\omega_f(b\alpha_o)[K_i\omega_i(b\alpha_o) + K_f\omega_f(b\alpha_o)]/b^2 \qquad (15)$$

Where we have introduced the notation $\omega_{i,f}^\dagger = \omega_i(b\alpha_o)\omega_f(b\alpha_o)$ to differentiate the value of the statistical weight function at the TS ($b\alpha_o$) factored by the symmetry parameter b. Similarly, for the hardness; Eqn 6 holds but we make the distinction on the location of the TS ($b\alpha'_o$) found through the hardness. With these considerations in mind, we write a symmetric activation ratio ($m_s^\dagger$) of the second derivatives as follows:

$$m_s^\dagger = \frac{[L_i\omega_i(b\alpha'_o) + L_f\omega_f(b\alpha'_o)]}{[K_i\omega_i(b\alpha_o) + K_f\omega_f(b\alpha_o)]} = \frac{\Delta\eta^\dagger[\omega_{i,f}^\dagger] - \Delta\eta \circ \beta'(b\alpha')}{\Delta E^\dagger[\omega_{i,f}^\dagger] - \Delta E \circ \beta(b\alpha_o)} \qquad (16)$$

Here the different values of have been considered. However, and because: $\alpha'_o - \alpha_o \approx 0$, their different value will not affect m_s. Consequently, both can be used regardless.

Equation (14) is of particular interest, and different to the one shown in Equation (6), and in the literature. The second local derivatives are here pondered by the statistical weight functions, which are not the same for all states, consequently they do not factor out among themselves. Introducing Equation (8) we write

$$\Delta \eta^\dagger[\omega_{i,f}^\dagger] = m_s^\dagger \Delta E^\dagger[\omega_{i,f}^\dagger] - \{m_s^\dagger \beta(b\alpha_o) - m_a \beta'(b\alpha_o)\}\Delta E^\circ \qquad (17)$$

This is an important result as we related $\Delta \eta^\dagger$ to $\Delta E^\dagger$ and ΔE°, all modulated by the second derivatives ratio ($m_s^\dagger$). Zhou and Parr [10] obtained a similar result: $\Delta \eta^\dagger = -\Delta E^\dagger/2$, that appears to be the average of the contributions. More recently, Toro-Labbe [2] and his collaborators obtained a rather more similar result. However, they do not include the symmetry factor in the treatment of the problem. The inclusion of the symmetry of the system under study through the parameter b in our treatment, is one of the more important contributions of this work as it has built in, as particular cases, results found in the literature.

The Absolute Transition State Hardness ($\eta^\dagger$)

In this section we want to find an analytical expression to compare the transition state hardness ($\eta^\dagger$) to the initial (η_i) and final (η_f) hardnesses. From Figure 14.1, we write

$$\Delta \eta^\dagger[\omega_{i,f}] = \eta^\dagger[\omega_{i,f}] - \eta_i[\omega_{i,f}] \quad \text{and} \quad \Delta \eta^\circ = \eta_f[\omega_{i,f}] - \eta_i[\omega_{i,f}] \qquad (18)$$

Replacing these relations in Eqn. (17):

$$\eta^\dagger[\omega_{i,f}] = m_s^\dagger \Delta E^\dagger[\omega_{i,f}^\dagger] - m_s^\dagger \beta(b\alpha_o)\Delta E^\circ + \beta'(b\alpha_o')\eta_f + [1 - \beta(b\alpha_o)]\eta_i \qquad (19)$$

Also, introducing Equation (18) into Equation (19), we find an equivalent relation:

$$\eta^\dagger[\omega_{i,f}] = m_s^\dagger \Delta E^\dagger[\omega_{i,f}^\dagger] - \{m_s^\dagger \beta(b\alpha_o) - m_a \beta'(b\alpha_o')\}\Delta E^\circ + \eta_i \qquad (20)$$

From here and Equation (16) we can re-write m_s at the Transition State ($m_s^\dagger$) as:

$$m_s^\dagger = \frac{\Delta \eta^\dagger[\omega_{i,f}^\dagger] - \Delta \eta^\circ \beta'(b\alpha_o')}{\Delta E^\dagger[\omega_{i,f}^\dagger] - \Delta E^\circ \beta(b\alpha_o)} \qquad (21)$$

This is an important result, because we have now related transition state parameters to initial and final conformations (local states). We shall return to these relation and their relevance, when we discuss numerical results.

Hammond's Postulate and The Brönsted Coefficient

When discussing the TS geometry, Hammond's postulate is often recalled [1-11, 12], to establish the resemblance of the transition state structure to reactants (exothermic reactions) or products (endothermic reactions). Simple energy comparisons are not sufficient to establish tendencies on reactions, and although the second derivatives at the reference conformations helps to establish trends, the relations derived for the hardness allows for more definite conclusions. In a previous work [1] we have derived an analytical expression for the Brönsted coefficient. In this respect, additional information is provided by this coefficient, which we claim to be a semi-quantitative expression for Hammond's postulate, as it provides the necessary insight, depending on the symmetry of the system under consideration [1-3]. This is in agreement with Leffler's postulate [12].

We have previously noted [1] that it is in fact the magnitude of the second derivatives of the potential function at the reference conformations with their respective statistical weights (see equations [1] and [2]), that really gives form to Hammond's postulate. In this work, we show additional evidence to support these findings with the relations found to represent the hardness as a function of the potential energy.

As discussed above, the Brönsted coefficient (β(bα)) suggests that the properties of the initial conformation (say trans), will insinuate the form of the final structure (cis). This becomes clear when we introduce the relations for the statistical weights of the reference conformations ω_i and ω_f as shown in Equations (2-a and 2-b). The evolution of the Bronsted Coefficient and the statistical weight function of the initial state are shown in Figures 14.3 and 14.4, respectively. From them it becomes clear that these functions are not the same for any system, as they clearly are a function of the symmetry properties of the system, as they should.

This suggests that the initial statistical weight function can be chosen as reaction coordinate [2], as it represents the relative distribution and evolution of the reference state along the reaction coordinate, as it varies from 0 (initial conformation) to 1 (final conformation) according to (2b).

Special Cases

a) Particular Case of b = 1:   We now examine the case of b = 1. This yields for the potential energy and the hardness functions the following relations ($b = 1$ in equations 1 and 5) for PEF and HF:

$$E(\alpha) = \Delta E° \beta(\alpha) + 2\omega_i(\alpha)\omega_f(\alpha)[K_i\omega_i(\alpha) + K_f\omega_f(\alpha)], \qquad (22)$$

and,

$$\eta[\omega_{i,f}] = \Delta\eta° \beta'(\alpha) + 2\omega_i(\alpha)\omega_f(\alpha)[L_i\omega_i(\alpha) + L_f\omega_f(\alpha)] \qquad (23)$$

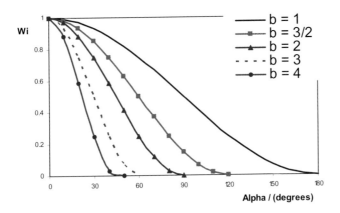

Figure 14.3. Statistical weight function $\omega_{if}(b\alpha)$ as a function of the dihedral angle (α) and the symmetry coefficient b.

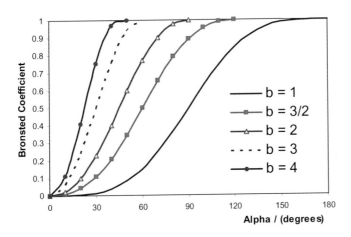

Figure 14.4. Conformational dependence of the Bronsted coefficient $\beta(b\alpha)$ with the symmetry coefficient and the dihedral angle α.

With this derivation we show that by examining this particular case, we re-derive results published previously for the potential [13] and the hardness [2]. This case will be widely studied in this work when we apply the procedure described here to a series of molecular systems.

b) The Symmetric Case ($\Delta E^\circ = 0$ and $\Delta \eta^\circ = 0$): For this case, the asymmetric contributions (Equations (3-b, 8 and 10-a)) are null. This is:

$E_a[\omega_{i,f}] = 0$; $m_a = 0$; $\Gamma_a[\omega_i(b\alpha),\omega_f(b\alpha)] = 0$

For the second derivatives, for both the potential energy and the hardness functions, we write:

$$K_i = K_f = \mathbf{K}; \quad L_i = L_f = \mathbf{L}; \quad m_s = \mathbf{L/K}$$

Therefore, we found simplified expressions for PEF and HF:

$$E[\omega_{i,f}] = E_s(b\alpha) = \omega_i(b\alpha), \quad \omega_f(b\alpha)\mathbf{K}/2b^2 = \mathbf{K}sin^2(b\alpha)/2b^2,$$

and,

$$\eta[\omega_{i,f}] = \eta_s(b\alpha) = \omega_i(b\alpha), \quad \omega_f(b\alpha)\mathbf{L}/2b^2 = \mathbf{L}sin^2(b\alpha)/2b^2$$

Some additional simplifications appear under the iso-energetic case ($\Delta E° = 0 = \Delta\eta°$). From Equations (2a and 2b) we find that at the TS (α_o):

$$\beta(b\alpha_o) = 1/2 = \omega_i(b\alpha_o) = \omega_f(b\alpha_o)$$

$$\beta'(b\alpha_o) = 1/2 = \omega_i(b\alpha_o) = \omega_f(b\alpha_o)$$

As a consequence, the transition state is easily characterized as the expressions for the energy and the hardness become:

$$E^\dagger[\omega_{i,f}] = E_s^\dagger(b\alpha) = \mathbf{K}/2b^2, \tag{24-a}$$

and,

$$\eta^\dagger[\omega_{i,f}] = \eta_s^\dagger(b\alpha) = \mathbf{L}/2b^2 \tag{24-b}$$

We note that the change in the symmetric potential and hardness functions changes according to the inverse of the square of b, decreasing monotonically as the systems move to higher orders of symmetry. This suggests that for high symmetry systems (b large), the value of the PEF or HF, or any other property represented through this algorithm will account for the stability of the system, as they converge to the asymptotic minimum value of zero (lim (b$\to \infty$) $1/2b^2$), stabilizing the system, as shown in Figure 14.5.

In addition, for the cases in which $\Delta E° = 0$ and/or $\Delta\eta° = 0$ the computations are reduced by a half because the potential energy function depends mainly on the value of the second derivatives which now can be computed at either the initial or final state, hence requiring only two points.

c) A Particular Symmetric Case ($\Delta E° = 0$, $\Delta\eta° = 0$ and $b = 1$): To examine this case, we introduce these constraints in Equation (7), and find the following:

$\eta[\omega_{I,f}] = \Gamma_s[\omega_{i,f}] = m_s[\omega_{i,f}(b\alpha)]\, E_s(b\alpha) = L/K$

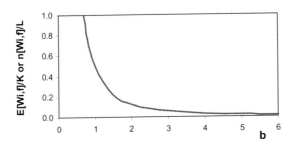

Figure 14.5. Change in the symmetric potential and hardness as a function of b according to Equations (24-a) and (24-b). These properties decrease monotonically as the systems move to higher orders of symmetry, as they converge to the asymptotic minimum value of zero ($\lim_{b \to \infty} 1/2b^2$), stabilizing the system.

Also,

$$E[\omega_{i,f}] = E_s(b\alpha) = K/2 \text{ and } \eta[\omega_{i,f}] = \eta_s(b\alpha) = L/2$$

From these relations, it is immediate that the TS (α_o) is at the midpoint between the initial and final conformations. By taking the first derivative of $E[\omega_{i,f}]$ and $\eta[\omega_{i,f}]$ we found $\alpha_o = \pi/2b$. Consequently, for b = 1 → $\alpha_o = \pi/2$. The same applies for higher values of b. The situation is particularly interesting because it reduces the number of points to be evaluated to two (initial or final states). In addition, from Eqn. (16):

$$\Delta\eta^{\dagger}/\Delta E^{\dagger} = L/K , \forall K \neq 0$$

3. Applications

There is a great amount of interesting examples are available in the literature to which this general procedure can be applied. Since the algorithm is particularly appropriate for the study of internal rotations, we include here some examples to show the usefulness of the procedure. For all the cases considered, calculations were performed by fully optimizing all the conformations considered at the Hartree-Fock-*ab initio* level. We have used the Gaussian 98 suite of programs [15], with a 6-31G basis set, with polarization orbitals. We have normalized all cases such that their energy has been set to zero at the initial state ($\alpha_i = 0$ – trans isomer).

The potential energy function behavior, properties and symmetry dependence has been discussed before [1, 2, 11]. The results obtained in this work are summarized in Table 14.1. We note that in all cases the principle of maximum harness [7] is obeyed.

We have chosen nine molecular systems that had been the subject of extensive theoretical and experimental research, in two groups. that are representative of

Table 14.1. The Symmetry Parameter (**b**), input data K_i, K_f, $\Delta E°$ and L_i, L_f, $\Delta\eta°$ necessary to define $E[\omega_{i,f}]$ and $\eta[\omega_{i,f}]$. Properties to characterize the conformational dependence upon internal rotation and molecular hardness are shown. K_i, K_f, L_i, and L_f are in kcal/mol-rad². Energies and Hardnesses are in kcal/mol. All m_a, m_s, β's, ω_i and ω_f are unitless.

	HONS	HSNO	HSNS	FONO	HO-NO	HOOH	HOOF	HSSH	Ethane
b	1	1	1	1	1	1	1	1	3/2
K_i	23.37	18.19	26.00	15.73	19.03	-2.88	-10.39	-9.47	-9.47
K_f	29.13	21.49	27.83	21.26	23.17	-13.81	-17.03	-18.04	-9.47
$\Delta E°$	0.38	0.44	-0.15	-1.99	0.83	7.97	-0.44	2.53	0.00
L_i	-6.59	-12.04	-4.91	-22.23	-7.05	12.10	12.73	16.87	-0.17
L_f	-17.60	-45.68	-13.71	-46.81	-22.88	22.02	17.12	17.33	-0.17
$\Delta\eta°$	3.24	1.82	0.97	3.62	4.23	-9.43	1.46	-0.30	0.00
α_o	90.42°	90.64°	89.84°	87.08°	91.13°	61.50°	90.93°	84.73°	60.00°
α'_o	82.30°	88.19°	87.03°	87.00°	81.88°	73.97°	92.80°	89.95°	60.00°
ΔE^+	13.33	10.15	13.38	8.23	10.99	-0.87	-7.11	-5.53	2.98
$\Delta\eta^+$	-4.28	-13.28	-4.08	-15.22	-5.11	4.44	8.28	8.53	-0.04
β	0.51	0.51	0.50	0.46	0.51	0.17	0.51	0.43	0.50
β'	0.40	0.48	0.46	0.46	0.39	0.30	0.54	0.50	0.50
ω_i	0.50	0.49	0.50	0.53	0.49	0.74	0.49	0.50	0.50
ω'_i	0.57	0.52	0.53	0.53	0.57	0.64	0.48	0.50	0.50
m_a	8.45	4.12	-6.36	-1.81	5.10	-1.18	-3.28	-0.01	0.00
m_{s-i}	-0.28	-0.66	-0.19	-1.41	-0.37	-4.20	-1.22	-1.78	
m_{s-f}	-0.60	-2.13	-0.49	-2.20	-0.99	-1.59	-1.01	-0.96	
m_{s-avg}	-0.44	-1.39	-0.34	-1.81	-0.68	-2.90	-1.12	-1.37	0.00
m_s^+	-0.47	-1.47	-0.35	-1.85	-0.71	-2.56	-1.09	-1.28	-0.01

many other bigger systems. Eight of them undergo internal rotation with the same point symmetry group, this is, $b = 1$, and $\alpha_f = \pi$. We have separated the molecular systems under study, The first group: HONS[16], HSNO[16], HSNS[2,17], FONO[18] (no polarization orbitals were used in the calculations for this system), and HONO[17,19], is represented by systems whose potential energy function describes a rotational isomerization that has the shape of a double-well potential, as is shown in Figure 14.6. Their hardness profile has a single well shape, and an opposite behavior to the PEF is displayed in Figure 14.7.

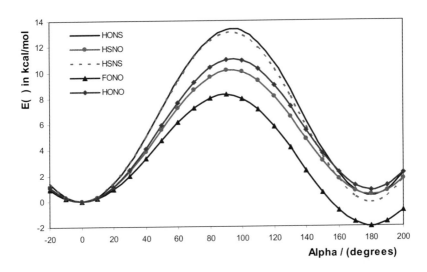

Figure 14.6. Potential energy surfaces for the internal rotation of the first group of molecules: HONS, HSNO, HSNS, FONO, and HONO systems. Initial and final states are located at $\alpha = 0°$ and $\alpha = 180°$ respectively. The TS (α_o) is located around midway between the initial and final states as shown in Table 14.1.

The second group: HOOH[20], (polarization orbitals - 6-31G** - were used for this system), HOOF[21], and HSSH[11,19,22] have a PEF that describes a *double-barrier* potential delineated in Figure 14.8. Their hardness function though, has the shape of a double-well potential, with an opposite behavior to the PEF as shown in Figure 14.9.

Finally, in Figure 14.10 the energy and hardness profiles are shown for the internal rotation of ethane, our last molecular example, which was chosen because of its characteristic form of being a simple hydrocarbon that well represents the structure and reactivity of many others.

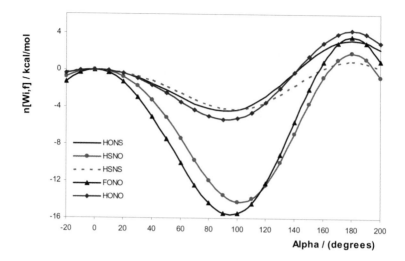

Figure 14.7. Hardness profile for HONS, HSNO, HSNS, FONO, and HONO systems, that have the shape of a single-well potential.

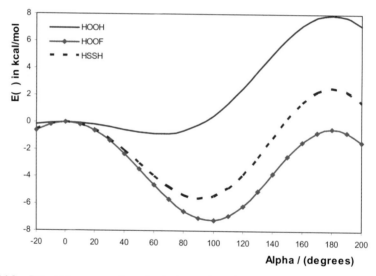

Figure 14.8. Potential energy surfaces for the internal rotation of the second group of molecules: HOOH, HOOF, and HSSH systems. Initial and final states are located at $\alpha = 0$? and $\alpha = 180$?, respectively. The TS (α_o) is located around midway between the initial and final states as shown in Table 14.1.

The $b = 1$ Cases: HONS, HSNO, HSNS, FONO, HO-NO, HOOH, HOOF, HSSH

For the first group we note from Figure 14.6 that if we consider as reference the HONO system, then most of the systems present an iso-thermic behavior, except for FONO whose cis isomer become more stable as its energy is lower. Although HSNO presents the same behavior on a lower energy barrier, it stays pretty much iso-thermic. On the other hand the inclusion of a sulfur atom in the vicinity of the nitrogen atom in HONS (but not connected to the hydrogen), appears to increase the energy barrier by more than 3 kcal/mol. Amazingly enough, the cis isomer of this specie (final state) has not yet been isolated. The hardness profile in Figure 14.7 confirms that FONO and HSNO are more reactive, this is, they have a lower energy barrier but deeper hardness when compared to HONO. The same behavior is observed for HONS and HSNS as they have the lowest hardness well making them less reactive, with more stable products (initial and final states).

The second group presents a potential energy profile, displayed in Figure 14.8, that is characterized by having the initial and final states as being transition states of a more stable conformation, the expected TS. This is an interesting behavior, which suggests that the hydrogen peroxide molecule is in equilibrium with its trans and gauche conformations. Hydrogen persulfide has a more stable TS (5.5 kcal/mol), as well as a more stable cis conformer (6.5 kcal/mol) than HOOH. On the other hand, HOOF has a more dramatic energy difference among its isomers, showing that its gauche conformation is more stable than the initial (trans) and final (cis) isomers by more than 7.0 kcal/mol and 0.5 kcal/mol, respectively. The evident effect of sulfur as central atoms and fluor in stabilizing the molecular system is confirmed by the hardness profile shown in Figure 14.9. In this case a deep well is shown for the trans conforma-

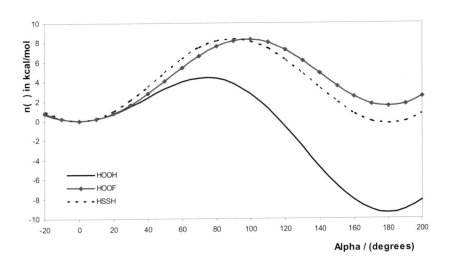

Figure 14.9. Hardness profile for HOOH, HOOF, and HSSH systems, that have the shape of a double–well potential.

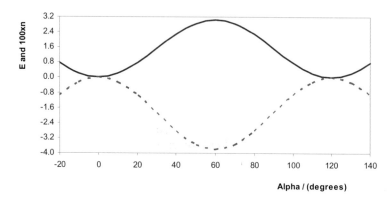

Figure 14.10. EthaneTorsional potential energy function (-) and hardness (——) profiles along α.

tion of the cis HOOH isomer indicating that the trans or gauche conformers are the more stable ones, having higher hardness barriers – less reactive; HOOF and hydrogen persulfide.

Ethane, a Case with b = 3/2

We choose this system because it is representative of a tremendous variety of molecular systems. Ethane's torsional energy barrier is well known [1]. We performed calculations following the staggered-eclipsed-staggered mechanism of this D_{3d}, i.e., $\boldsymbol{b} = 3/2$, with $\alpha_f = 2\pi/3$, with the saddle point located at $\alpha = 60°$. The torsional barrier of ethane is very well known, and reported theoretical and experimental values oscillate between 2.7 and 3.0 kcal/mol [22], depending on the level of theory used. In this study we report a barrrier of 2.98 kcal/mol.. This system falls in the special case discussed before, this is, $\Delta E° = 0$ and $\Delta\eta° = 0$, with b = 1. The input parameters and relevant data are displayed in Table 14.1.

In Figure 14.10 the potential energy and hardness profiles for the internal rotation of ethane are shown. The hardness function has been amplified by a factor of 100 to show its form and evolution along the reaction coordinate (α). The lower values of the hardness along α indicates that this system undergo internal rotation, with a low barrier to be surpassed.

For the Bronsted coefficient (β) and the initial state statistical weight function (ω_i) we have computed its the values for PEF (β and ω_i) and HF (β' and ω'_i), which are significantly different, and for whom there is no simple relation to write one as a function of the other. Results shown in Table 14.1. The value of m_a does not provide additional insight on the form of the potential and hardness function. It is indeed the ratio of the forces (m_s) that gives more information regarding the form of the isomerization surfaces.

For the ratio of the forces (m_s) four sets of values are displayed in Table 14.1. The first 2 (m_{s-i} and m_{s-f}) are those of the initial and final state according to Equation (16). An intermediate value (m_{s-avg}) computed as the average of the initial and final values, is also listed. Finally, the value of m_s at the transition state $m_s^\dagger$, which was computed using the values of α_o, is shown at the end of Table 14.1. We note that both the average and the transitions state computed values are not too different from each other, suggesting that the average value could be used, although the variation of the forces ratio with the reaction coordinate and the symmetry of the system suggests the contrary. In addition, when we calculated m_s at a value of α which is midway between the initial and final states considered ($\alpha = 90°$ for the 8 first molecules, and $\alpha = 60°$ for ethane), the difference is very small. Finally, we also computed the value of the force ratio using the different values of the reaction coordinate at the TS for the PEF and the HF (α_o and α'_o, respectively). We found that there was no difference between this value and the reported figure of $m_s^\dagger$ shown in Table refc2t1.

4. Conclusions

A symmetry-adapted interpolation procedure to find transition states, and to represent the hardness function $\eta[\omega_{i,f}]$ in terms of the potential energy function $E[\omega_{i,f}]$ in molecular systems that undergo internal rotations have been presented. These global functions are written in terms of the statistical weight functions $\omega_{i,f}$, which in turn are functions of the principal symmetry axis of the point group of the molecular system under consideration, and the torsional angle (α) (reaction coordinate. This results in a substantial reduction in the computational efforts (about 80%) to be performed to study the potential energy and hardness surfaces.

The features of the procedure have allowed us to reproduce already existing algorithms, and to show that they are particular cases of this more general symmetry-adapted method. As a consequence we derived general relations to correlate the HF with the PEF. Also, we proof that the forces (second derivatives) ratio is indeed a function of the symmetry (b) of the system and the reaction coordinate (α) (Figure 14.2). On the other hand, we also showed that the Bronsted coefficient and the statistical weight functions that correlates the PMH with Hammond's postulate, are a function of both α and b, as shown in Figures refc2f4 and 14.5. They also give us insight on the influence of the forces on the location of the TS, suggesting the PEF barrier, and the resistance of the system to undergo internal rotation, this is, its hardness.

In our study we have considered hypothetical cases for different values of b and apply those relations to nine molecular systems. These special cases gave us more insight on the variables governing the internal rotation of the system under given conditions. This was clearly established when we studied the symmetric case and write an expression for both the PEF and the HF's, and show that they asymptotically converge to a given value, as b increases, suggesting that higher symmetry systems are indeed more stable.

Our numerical results seem to be in good agreement with those found in the literature (see for example reference [2]). We have shown that theoretical expressions currently available in the literature are particular cases of the general symmetry-adapted interpolation procedure shown here.

In an initial stage, one of the goals of our research was to show the ability of the algorithm to give a good estimation of the location of the TS. In this work we are concern with the ability of the algorithm to reproduce the hardness profile and find its minimum

A correlation between the molecular Hardness and the $E(b\alpha)$ is shown, provided that their dependence on α can be described by the algorithm shown in this work.. Also, the activation Hardness is written in terms of the activation energy, and the energy and hardness difference of stable isomers.

We have not make an attempt to obtain optimized parameters [2], as the algorithm has shown its accuracy with a minimum number of input data, i.e., two in each state (initial and final). This kind of extra efforts are useful to ensure the reliability of the procedure. However, when we compare with results available in the literature, we notice that this extra effort is not required by our procedure, mainly because of its symmetry adaptation, which plays a key role in the expression of the forces ratio. Finally, the ratio of the forces shows that there is no specific value to be used, but that we should rather use the weighted forces expression as prescribed by equation (6)

The technique can be used to correlate any global property with the potential energy, represented by the same analytical form (Eqn. (1)). Moreover, and as shown in a previous work [1], the procedure accuracy is not a function of the level of theory used.

Acknowledgments

The author would like to express his gratitude to Professor Panos Pardalos from the Industrial and Systems Engineering Department at the University of Florida for his support and encouragement.

References

[1] C. Cardenas-Lailhacar and M.C. Zerner, *Int. J. Quantum Chem.* 1999, **75**, 563

[2] Cardenas-Jiron, G.I., Lahsen, J., Toro-Labbe, A. *J. Phys. Chem.* 1995, **99**, 5325; Cardenas-Jiron, Toro-Labbe, A. *J. Phys. Chem.* 1995, **99**, 12730.

[3] Parr, R.G.; Pearson, R.G., *J. Am. Chem. Soc.* 1983, **105**, 7512.

[4] Pearson, R.G., *J. Am. Chem. Soc.* 1985, **107**, 6801.

[5] Parr, R.G.; Yang, W. *Density Functional Theory of Atoms and Molecules*; Oxford University Press, New York, 1989.

[6] Parr, R.G.; Zhou, Z. *Acc. Chem. Res.* 1993, **26**, 256.

[7] Parr, R.G.; Chattaraj, P.K. *J. Am. Chem. Soc.* 1991, **131**, 1854; Pearson, R.G., Palke, W.E. J. *Phys. Chem.* 1988, **180**, 209.

[8] Parr, R.G.; Palke, W.E. J. *Phys. Chem.* 1988, **180**, 209.

[9] Pearson, R.G. *Chemical Hardness;* Wiley-VCH, Weinheim – Germany, 1997.

[10] Zhou, Z. Parr, R.G. *J. Am. Chem. Soc.* 1990, **112**, 5720.

[11] Cardenas-Jiron, G.I. , Cardenas-Lailhacar, C. and Toro-Labbe, A., THEOCHEM 1993, **282**, 113; Chattaraj, P.K., Naith, K., Sannigrahi, A.B., *J. Phys. Chem*, 1994, **98**, 9143.

[12] Hammond, G.S., *J. Am. Chem. Soc.* 1955, **77**, 334.

[13] Leffler, J. E., *Science* 1953, **117**, 340.

[14] Cardenas-Jiron, G.I. , Cardenas-Lailhacar, C. and Toro-Labbe, A., *J. Mol. Structure (Theochem)* 1990, **210**, 279 ; Cardenas-Jiron, G.I.and Toro-Labbe, A., Chem. Phys Lett. 1994, **222**, 8 ; Cardenas-Jiron, G.I., Toro-Labbe, A., Bock, C.W. and Maruani, J. In *Structure and Dynamics of*

Non-Rigid Molecular Systems; Smeyers, Y.G., Ed., Kluwer Academic: Netherlands, 1995, pp. 97- 120.

[15] Gaussian 98, Frisch, M.J., Trucks, G.W., Schlegel, H.B., Gill, P.M.W., Johnson, B.G., Robb, M.A., Cheeseman, J.R., Keith, T. Petersson, Montgomery, J.A., Raghavachari, K., Al-Laham, M.A., Zarkrzewski, V.G., Ortiz, J.V., Foresman, J.B., Cioslowski, J., Stefanov, B.B., Nanayakkara, A., Challacombe, M., peng, P.Y., Ayala, P.Y., Chen, W., Wong, M.W., Andres, J.L., Replogle, E.S., Gomperts, Martin, R.L., Fox, D.J., Binkley, J.S., Defrees, D.J., Baker, J., Stewart, J.P., Head-Gordon, M., Gonzalez, C., and Pople, J.A., Gaussian, Inc., Pittsburgh PA, 1998.

[16] Nonella, M., Huber, J.R and Ha, T.K. *J. Phys. Chem.* 1987, **91**, 5203 ; Muller, R.P., Nonella, M., Russeger, P. and Huber, J.R., *Chem. Phys.*, 1984, **87**, 351.

[17] Cardenas-Jiron, G.I., Letelier, J. R., Maruani, J. and Toro-Labbe, A., Mol. Eng., 1992, **2**, 17.

[18] Cardenas-Jiron, G.I. and Toro-Labbe, A., *Chem. Phys. Lett.* 1994, **222**, 8 ; Dixon, D.A. and Chritie, K.O, *J. Phys. Chem.*, 1992, **96**, 1018.

[19] Toro-Labbe, A., THEOCHEM 1988, **180**, 209 ; Cardenas-Jiron, G.I. , Cardenas-Lailhacar, C. and Toro-Labbe, A., THEOCHEM 1990, **210**, 279 ; Cardenas-Jiron, G.I. and Toro-Labbe, A., An. Quim. (Madrid), 1992, **88**, 43; Darsey, J.A., Thompson, D.L., *J. Phys. Chem.*, 1987, **91**, 3168 ; Letelier, J.R, Toro-Labbe, A. and Utreras-Diaz, C., *Spectrochimica Acta*, 1991, **A47**, 29 ; Cox, A.P., Brittain, A.H. and Finnigan, D.J, *J. Chem. Soc. Faraday Trans.*, 1971, **61**, 2179 ; McGraw, G.E., Bernitt, D.L and Hisatune, I.C., *J. Chem. Phys.*, 1966, **45**, 1392 ; Deeley, C.M. and Mills, I.M/, *Mol. Phys*, 1985, **54**, 23.

[20] Hehre, W.J., Radom, L., Schleyer, P. v. R. and Pople, J.A., *Ab-Initio Molecular Orbital Theory*; John Wiley & Sons: New York, 1986; p. 267. ; Hunt, R.H., Leacock, R.A., Peters, C.W. and Hecht, K.T., *J. Chem. Phys*, 1965, **42**, 1931.

[21] Francisco, J.S., *J. Chem. Phys*, 1993, **98**, 2198.

[22] Herbst, E. and Winnewisser, G, *Chem. Phys. Lett.*, 1989, **155**, 572 ; Herbst, E. Winnewisser, G, Yamada, K.M.T., DeFrees, D.J. and McLean, A.D., *J. Chem. Phys.*, 1989, **91**, 5905; Cardenas- Lailhacar, C. and Toro-Labbe, A., *Theor. Chim. Acta*, 1990, **76**, 411.